AF869076

LOLA's FIRST: ***The Mark 1***

56
56

LOLA's FIRST:

The Definitive History of the *Mark 1*

DAVID PRATLEY

BOOKMARQUE
PUBLISHING

Minster Lovell & New Yatt • Oxfordshire

First published 1998

© David Pratley 1998

The author asserts his moral rights

All rights are reserved.
No part of this publication may be reproduced, stored in a retrieval system or transmitted, in any form or by any means, electronic, mechanical, photocopying, recording or otherwise, without prior permission from the publisher or copyright holder.

British Library Cataloguing in Publication Data
A catalogue for this book is available from the British Library

ISBN 1-870519-47-7
LIMITED CASEBOUND EDITION

164

Frontispiece: The prototype at Crystal Palace 5th July 1958

Set in 11pt on 14pt Palatino
Typesetting and origination by Bookmarque
Printed on Fineblade 115gsm
Published by Bookmarque Publishing
Printed and bound in Great Britain

Contents

A thoughtful Eric Broadley, in period fisherman's knit sweater, working out ways to beat the Lotuses.

Preface

THIS BOOK is a small tribute to Eric Broadley, and to the car that he designed in 1958 which went into production as the first model of Lola Cars Ltd. At the time, it was simply known as the Lola sports car. With production of the first single seater in 1960, and then further models, it quickly became retrospectively christened the 'Mark 1'.

Everybody has a motor racing hero. Perhaps most people would name a great driver of the past but, as an engineer myself, I have always been more attracted to the car designers and builders. My own choice, therefore, is that modest ex-chartered surveyor who went on to found one of the world's most successful production racing car companies.

I attended my first race meeting at Goodwood in 1951. Like many a schoolboy before and since, I was immediately hooked on the sport, and proceeded to give myself regular adrenalin injections every Friday by sneaking a free look at the fledgling *Autosport* in Stephens' Bookshop, near Fleet Street.

In the 1950s I was at London University, reading Chemical Engineering, as, a couple of years earlier, was one B Ecclestone, although he made a more profitable career choice than I did! My first car was inevitably an Austin 7 Special—which someone else had made a start upon—and I subsequently turned it into a reasonably reliable student runabout. Such reliability was not without cost to my mother's lounge carpet, which bore forever the scars of several engine rebuilds. When I acquired the Special it was only natural to join the 750 Motor Club, even though I had no hope of being able to afford to compete myself. I did, however, pore over the monthly Club *Bulletins* to read about the exploits of Colin Chapman, Major Mallock and the others in that college of post-war British motor racing development. It was clear in 1957 that the Broadley cousins, with their one-off

Special, had started to take on the then all-conquering Lotus in the premier 1172 cc class with some startling results.

A year later, I found myself spectating at Crystal Palace on 5th July 1958 when the Broadleys began to repeat the process of challenging the Lotus establishment again, but at a higher level. The beautiful little silver car made a lifelong impression on me, and—some 40 years later—has led to this book. I hope you have as much enjoyment reading it as I have had over the years, both in researching the history, and racing one of these delightful cars myself.

David Pratley
Wentworth, Surrey

Acknowledgements

I PARTICULARLY owe a debt of gratitude to Graham Broadley, both for his memories of the period, and for contributing the Foreword at the very last moment, when Eric, understandably, had pressing business in Huntingdon. Equally, I could not have put pen to paper without major contributions from Rob Rushbrook MBE, Mike Hyatt, George Rance, Dick Berger (who unearthed the original files of Allan Ross), and Ted Walker of Ferret Fotographics, who provided many of the photographs from his extensive archives.

Additional invaluable contributions have been made by Peter Gammon, Peter Ashdown, Duncan Rabagliati, Dick Scott, Charlie Gibson, David McKay, David Bertram, Ian McDonald, David Grossman, Norbert Michel, Dave Friedman and Richard Pelatowski who provided some splendid period US photographs, Willy Widar, Harry Reynolds, Mike Peck of the 750 Motor Club, Alistair Caldwell, Jannie van Aswegen, Bobby Bell, Dr Tony Goodwin and Bernard Cowdrey.

Thanks are also due to John Rose of Bookmarque, who has been courageous enough to publish my ramblings; and to my wife, Joan, who has proof-read them and corrected my punctuation and my notoriously poor spelling. My sister-in-law, Mary Jagger, came to my rescue when my aged Amstrad blew up on the last lap!

Additionally, I have borrowed much excellent text from *Autosport*, *Motor Sport* and *Motor Racing Australia*, all of whose Publishing Directors have kindly given me permission to use extracts from their magazines. I apologise if I have inadvertently failed to acknowledge the odd photograph or article reference. It is certainly not for want of trying, but tracing ownership after periods of nearly 40 years is not always possible.

Foreword

I AM DELIGHTED to support David Pratley in writing this book about our earliest cars, not only because he has put a great deal of effort into the research, but also because his attitude and enthusiasm reflect perfectly those of the keenly enthusiastic 'racers' of the late '50s and early '60s.

You will see that we sold cars to people of all strata of society and that the racing was very 'amateur', even in international events, by comparison with today's standards. Design computers were some 20 years away; aerodynamics a foggy area; racing tyres at an early stage of development—wets, dries and intermediates a dream (or nightmare) for the future; pitch and roll sensitivity were ideas, but no more than that.

The leviathan of technology has rolled on, consigning this happy time for motor racing to the archives. I hope that this book will enable you to remember, if you are old enough, or imagine, if you are not, the friendliness, keen enthusiasm and social spread of this era.

GRAHAM BROADLEY
Sundridge, Kent
August 1997

Chapter 1

Start of the learning curve: The Ford Side-Valve Special

MOTOR RACING before the Second World War was mostly the province of the rich. The often quoted old Brooklands motto "the right crowd and no crowding" sums up the way things were done, with only the legendary Freddie Dixon doing some crowding and who certainly came from the wrong crowd. The war changed all that. It had introduced many of the servicemen to the skill of driving and, in the immediate post-war period, this created aspirations of vehicle ownership. It took time to turn the munitions programme into car production, and for a period virtually the only cars that could be purchased were pre-war. Even the limited post war production was to be of cars that had been designed six or more years before.

For the man-in-the-street, the extension of this enthusiasm for everyday motoring into some form of competition was little more than a dream. Fortunately mankind has always had a few dreamers who will not be deterred by lack of resources and, as far back as 1936, a group of enthusiasts in the Bristol area had borrowed from the more cosmopolitan world of motorcycling. They built a few simple cars for grass track racing, powered by motorcycle engines. At the end of hostilities they formed themselves into the "500 Club" to promote the idea of a do-it-yourself real racing car costing not much over £100. From this modest beginning the 500 movement took shape, with affordable Norton or JAP-powered racing cars that turned out to be a major proving ground for aspiring drivers and designers in the early and mid-1950s.

"A 750 Club – If Needed"

In 1939, the year production of the Austin Seven ceased, *Motor Sport* editor, Bill Boddy, wrote a piece in his magazine entitled "A 750 Club – If Needed". It ran:

> "In view of the fact that class divisions, which are becoming increasingly popular in trials, usually start at up to 1100 cc, the Austin Seven as a 747 cc side-valve car is apt to be outclassed in its older form as owned by a large body of enthusiasts. The obvious cure seems to be to found a Seven-Fifty Club, exclusively for Austin Seven owners, and run sensible trials for these fascinating babies, possibly permitting entries from such things as the inevitable Morris Cowley and the 850 cc M type MG on occasion. Quite an interesting one-make club should result, because so many variations of the Austin Seven exist, mostly of sports type body, while probably more conversions are based upon this chassis than any other."

Enough people thought this idea worthwhile, and the 750 Motor Club eventually came into being.

From trials to the track

Holland Birkett, who was Chairman of the 750 Motor Club, proposed a "750 Racing Formula" in 1949. Basically this was to be a cheap formula derived from an Austin Seven chassis, rear axle, engine block and crankcase, and gearbox. Choice of front axle and suspension were free. The body needed to conform to RAC regulations for trials and rallies, and had to have legal electrical equipment. Cars had to be driven on the road to events.

The first race for the new fledgling racing cars was at the Eight Clubs meeting at Silverstone on the 3rd June 1950. A decent seventeen cars entered for the 5 lap race, with the fastest cars starting at the back of the grid!

The event was won by Charles Bulmer driving Birkett's own car at a speed of 56.10 mph. Elsewhere on the programme, a young 20-year-old by the name of Colin Chapman with a Ford side-valve powered device called a Lotus Mark II had won a scratch race from Dudley Gahagan's Type-37 Bugatti.

... and on to the Ford engine

Following the success of racing based on the Austin Seven, there was pressure from within the Club to allow a more powerful form of competition based on the cheap, and readily available, Ford 1172 cc side-valve engine. An "1172 Formula" was proposed in May 1952—this was similar in concept to the 750 Formula. The power unit, of course, was initially based on the Ford Prefect, which itself was shortly to be discontinued, although the engine was retained in the Ford Popular. The standard engine had a bore and stroke of 63.5 by 92.5 mms. It managed less than 30 bhp at 4,000 rpm from a 6.2:1 compression ratio. A reasonable increase in power output for racing purposes could be achieved by applying typical tuning attention to porting, adding twin carburettors, changing valve timing and going up to 8.5:1 compression ratio with the help of a Ford Anglia head and 20 swg copper gasket. Supercharging, o.h.v. conversion, de-siamezing of inlet ports and reversal of port function was banned.

The Silverstone Eight Clubs meeting was again the event where the new formula made its debut. This time, however, on the 7th June 1952, only eleven entrants could be mustered, so they formed part of another scratch race which was won by future Northamptonshire Chief Constable John Gott's HRG.

The 1172 Formula section turned out to be a 1-2-3 for Buckler cars.

* * *

The Broadley cousins

Eric Broadley left Bromley High School at 17 and worked for a local building firm called Treasures. He continued his education at night school by studying building technology and quantity surveying. Transport, at that time, came from an old motorcycle, but Eric's main hobby was constructing model boats and aeroplanes. There were no handy plastic kits around in those days so modellers had to scratch-build their creations, although tiny dope-fuelled engines were beginning to be imported from the USA. Eric, however, had a model engineer's Myford ML7 lathe, and set about building his own 5 cc and 10 cc engines.

In 1954 his engineering interests had extended to full-sized machinery, and he joined the 750 Motor Club to construct an Austin 7 Special.

Cousin Graham, meanwhile, had joined the Royal Navy under the 'Y scheme' to be offered training as either a Sick-Bay Attendant or a Steward. Less than impressed with either of these career choices, he transferred to the Fleet Air Arm and promptly took a course in aircraft engine maintenance. Before long, at the ripe old age of 18, he found himself signing off on engine rebuilds!

On leaving the Service he joined the family outfitting business in Bromley as a gentlemen's tailor, but he, too, constructed an Austin 7 Special in his spare time by putting his Fleet Air Arm experience to good use.

Graham's father was strictly religious, and confined any sporting activities to Saturdays only; hence cousin Eric got to drive the car on Sundays and confessed, after one Silverstone meeting, to being "frightened to death."

* * *

The Ford Special

Clearly there had to be something better than Graham's potentially lethal Austin 7 Special. In 1956 the cousins decided that they really wanted to try the 1172 Formula, and that one of the new Lotus Mark VIs was just what was needed. They figured that if they sold both Austin 7s they could just scrape together enough for the Lotus. Come the day of reckoning, however, and two second-hand Specials did not add up to one new Lotus from young Mr Chapman's new company. So, it had to be Plan B — build one themselves.

Work started in the lock-up garage behind Broadley Bros (Outfitters) in late 1955 on the first car that would carry the Lola name.

The car was typical of the racing Specials of the period which had, by then, displaced the early road-derived Bucklers and Dellows. It featured the usual swing axle front suspension, live rear axle, and home-brewed Ford 1172 cc side valve engine. Significantly it was exceptionally well constructed; a feature

Eric Broadley in the 1172 cc Formula race at the 1956 'Eight Clubs' meeting at Silverstone..

(above) *A rare outing for cousin Graham in 1956.*

(right) *An evocative paddock shot taken at the 1957 'Eight Clubs' meeting. Note the first appearance of the Lola marque identification on the nose cowl.*

which became a characteristic of all subsequent cars to carry the Lola name. It did not set the 1172 race world on fire in 1956, but the cousins soon learned one of the first lessons in motor racing: there are some rules which can be interpreted more elastically than others. This applied particularly to what could be done to tuning the engine, and it was clear that the quick men had got some rather special tweaks which the Broadley motor had not.

Much thought went into the engine for 1957. For example, Eric pressed the Myford into service to make lightened concave bottomed tappets out of an old halt-shaft, after much calculation of valve lift velocity and acceleration. Similar painstaking effort went into the other engine components with the result that the Lola had many more horses for its second season.

* * *

1957: Lola hits the front

First outing of the new season was at the opening BRSCC meeting at Brands Hatch on the 31st March. There were seven Lotuses among the nine starters but, although it led for the first two laps, the Lola recorded a DNF.

Eric did rather better with a fourth place in the up to 2000 cc sports car race on Easter Monday, again at the nearby Kentish venue.

The Maidstone and Mid-Kent race meeting at Silverstone on Saturday 18th May was the first to be held after the removal of petrol rationing, brought about by the Suez crisis. It also was notable for a first Lola win.

The *Autosport* reporter commented:

> "It seems necessary to have an expensive motor car for success in these (1172 cc) races. The outstanding exception to this rule was G(sic) H Broadley, whose own creation, Lola, goes considerably faster than the Lotuses and brought him romping home with a substantial lead."

A week later at the Cheshire circuit of Oulton Park, Graham repeated the dose, this time beating off John Anstice-Brown's Lotus.

At the Eight Clubs' Silverstone meeting, Eric beat Dick Prior's Lotus in the 1172 Formula race, and was a close second to Anstice-Brown in a 5 lap handicap.

Again, *Autosport* was impressed:

> "Once again E H Broadley showed his own Lola was a match for the latest factory made Lotuses."

The Trio Club meeting at Brands Hatch was held for a period in the later 1950s and early 1960s with members from Club Lotus, the 750 Motor Club and the 250 MRC participating. Eric Broadley again showed the Lotuses a clean pair of heels in the 1172 Formula Final, winning at an average speed of 65.04 mph.

The following week found the Bromley boys in Ulster at the old airfield circuit, just south of Ards, known as Kirkistown. 1172 Formula racing had taken root on the other side of the Irish Sea, and the locals were keen to prove that their machines could

The whole town turns out to watch scrutineering of the English cars for the Irish Ford Championship at Kirkistown in June 1957.

more than sort out the English ones. The event was grandly named the Irish Ford Championship, although, in fact, the invaders took the first three places with Eric winning by some 14 seconds. After the race, he got talked into staying on in Ireland for some holiday and to run at the following weekend's Leinster Trophy meeting. The extension was worth it as Eric came second in the 2-litre unsupercharged class at the Wicklow circuit.

Later, this lengthened stay was to cause dramas with his employers back in Bromley for they had been expecting his return somewhat earlier!

Back on English soil, the Lola continued to be the class of the field in the 1172 Formula. A breathless *Autosport* reporter wrote of the SUNBAC Silverstone meeting in September:

> "1172 race was a great triumph for E H Broadley who fought and won a stirring duel with R N Prior's Lotus 11. Lola led, was passed, led again, passed again, shot each side of a tail-ender at Woodcote. Almost together. Lola by a wheel."

Two further Silverstone wins were recorded by Eric at the Peterborough and Northampton Motor Clubs, and the North Staffs meetings, each time beating off the faster Lotus Elevens.

The 1957 season had been so successful that Eric won the major trophy for the 1172 Formula cars, presented to the 750 Motor Club by none other than Colin Chapman.

It was the first time that Lola had tweaked Lotus's tail — it would not be the last.

CHAPTER 2

The Prototype

TOWARDS THE end of the very successful 1957 season, Eric and Graham began to plan for the following year. Graduation to nationally recognised small sports car racing was the next obvious, if challenging, step. Colin Chapman had followed this route in 1954 with the Mark VIII, and then the Mark IX a year later, and by 1957 was dominating this category with the Series 1 Lotus Eleven both nationally and internationally. Furthermore, the upcoming young drivers were increasingly seeing this class of sports car racing as a career step, rather than the 500 cc F3 single- seaters that had produced the likes of Moss, Collins, and Lewis-Evans, five years before.

"The fire pump that wins races"

The key to this popularity of under 1100 cc and 1500 cc sports car racing was the Coventry Climax engine. Before this became available in 1954, the British competition had mainly relied upon tuned MG XPAG engines. However, from the unlikely combination of the Korean War and a fork-lift truck manufacturer, emerged a fire pump engine that formed the basis of one of the classic small racing engines.

The Company's roots went back to 1903 when H Pelham Lee came back from the Boer War and set up a car-making business in Coventry with a Dane by the name of Stroyer. The Lee Stroyer did not make much of a ripple in the motor carriage sales of the time, and Stroyer soon left the Company. Lee was convinced that there was potential for him somewhere in this new industry, and decided to concentrate his effort on engines. The resulting Coventry Simplex was a 1389 cc four-cylinder side-valve unit of 66.5 mm x 100 mm bore and stroke. The head and block were cast integrally. The engine found its way into a number of marques, most notably the Isotta-Fraschini, originally built for the 1908 Coupe des Voiturettes at Dieppe, and the very first Lionel Martin-built Aston-Martin in 1914. It was also used for such diverse activities as First World War searchlight generators, boats, and the snow tractors on Shackleton's Antartic expedition. Such versatility would become apparent in another of the company's products 40 years later.

Lee changed the company's name to Coventry Climax in 1917 and built engines for light cars such as Crossley, Morgan, Swift, Triumph and others. As the car market contracted in the 1930s, Lee's son, Leonard, began building engines for Government generator and pump contractors, and later started to manufacture the whole machines. With a design borrowed from the Hale Fire Pump Company of America, Coventry Climax provided excellent support to the war effort during the 1939-1945 conflict.

After the war, Leonard Lee turned to building fork-lift trucks. However, the Korean War in 1950 re-energised his Government work. The old fire pump engine was a 20 hp, 800 cc, cast-iron side-valve unit which dated back to the Swift days. The new Home Office specification called for a pump driver with half the weight and twice the power of the old engine. Design for the new unit was the responsibility of Walter Hassan, who had recently been appointed as Chief Engineer, and had a curriculum vitae that included Bentley, ERA and Jaguar. Perhaps appropriately, in view of Graham Broadley's background, he was the son of a Hackney outfitter of Saracen descent.

Hassan worked with Harry Mundy, who was an old friend from Bentley and ERA days, and Harry Weslake, who had worked on the Jaguar XK 120 head and did the gas flow, to lay out an engine that had similar requirements to a racing engine: something light and efficient and giving the best possible power to weight ratio.

The resulting design was a four cylinder, single overhead camshaft, engine of 1020 cc. It had an aluminium head and block with steel liners. The valves ran in wedge-shaped combustion chambers, and with inverted bucket and shim valve gear similar to the XK Jaguar. When first cranked up on a Saturday afternoon in April 1951, it gave 2 bhp more than the specified 35 bhp at 3,500 rpm on the dynamometer. They called it the Climax FW for 'Featherweight' — it weighed in at just under 200 lbs. The Home Office were properly delighted at the performance of the new engine, and placed an order for over 5,000 units.

As interest grew in small sports car racing, some of the manufacturers that had used MG engines — Kieft, Lotus, and Cooper — noticed the potential of the Climax FW, and badgered Lee to turn the FW into a motor car engine. The fire pump became transmuted into the FWA — 'Featherweight Automotive'.

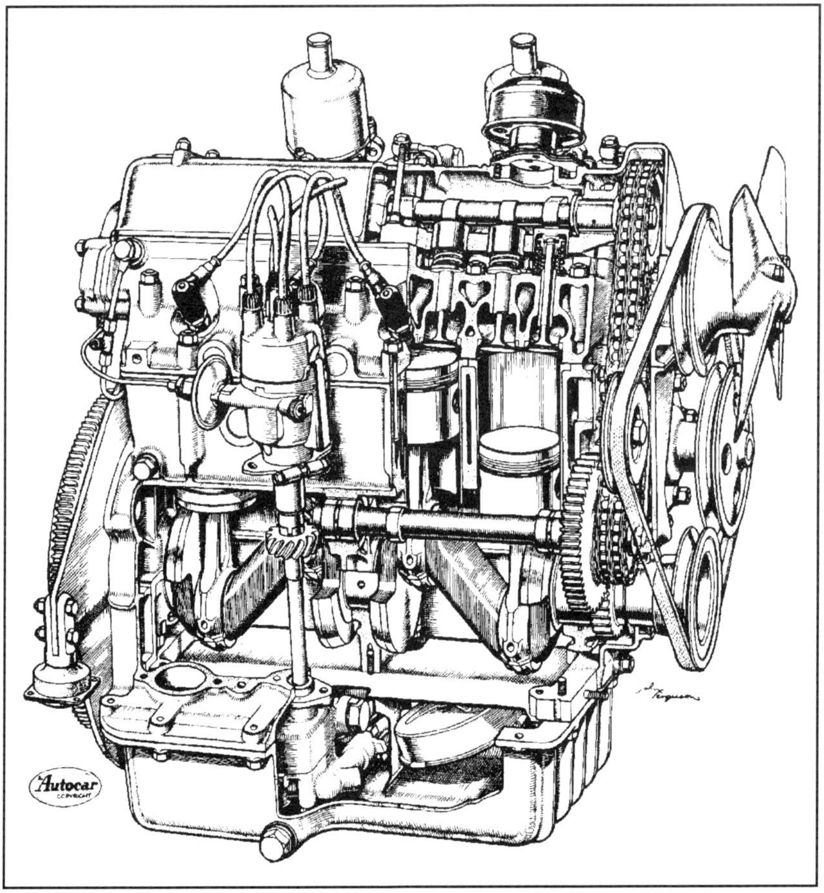

Cutaway drawing of the Coventry Climax FWA.

The FWA had a slightly longer stroke (66.7 mm vs 62.5 mm) than the fire pump engine, and a bore increased by 0.4 mm to 72.4 mm to give a capacity of 1098 cc. The compression ratio was 8.8:1, reworking the porting and cam timing, upping the oil pump capacity by 40%, and replacing the downdraught carburettor with a couple of SUs, resulted in a healthy 72 bhp at 6100 rpm. The cast-iron fire pump crankshaft was changed for a forged-steel unit. Maximum torque was 70 ft/lb at 4,750 rpm. The first few units were available for customers in 1954, and were your's for £250.

As *Motor Racing* remarked at the time:

"These days quite a number of people know how to build roadworthy chassis... It is the engine which is the most difficult to produce, and the appearance of a new light and powerful engine on the market is the signal for designers to start work on new cars to make use of it."

Lady Godiva rides the cambox cover of all FW engines.

The first FWA to race went straight into the Alan Rippon/Bill Black Kieft for the 1954 Le Mans 24 Hour Race. A testing debut! It was obvious to Harry Mundy (who was sent to Sarthe by Climax) as soon as the car ran on the circuit, that some baffling in the sump would be required. This cured the worst of the oil surge, and the engine ran like a sewing machine for 10 hours — until the car's back axle failed.

A second Kieft-Climax ran later that year in the TT at Dundrod on 11th September, and scored the FWA's first class win.

Climax, meanwhile, further developed the engine and, by increasing compression to 9.8:1 and changes to cam timing and intake manifolding, got 75 bhp at 6500 rpm. They then pushed the compression up again to 10.5:1 for so called Stage 3 tuning. This level also required new pistons with valve recesses machined in the crown, steel timing gears, a five bearing carrier for the camshaft rather than three, 9 mm connecting-rod bolts, and stronger valve springs. When all done, Stage 3 tuning gave over 90 bhp, and a shade more with Weber carburettors. As Walter Hassan recalled in his book *Climax in Coventry* (Motor Racing Publications), this did not stop the customers trying their own tweaks: Colin Chapman sent a unit back to Coventry for testing before a race. It blew up on the dynamometer, and when Climax examined the remains they found there were all sorts of modifications including an astronomic compression ratio. They returned the unit to Chapman as it had finished its test — in pieces!

Ultimately, well over 2,000 of the basic FW engine were built for motor vehicle use in various forms, of which 697 were 1100 cc 'racing fire pump' engines. The bore and stroke were also later increased to 76.2 mm and 80.0 mm to produce the 1460 cc FWB. By using the FWB pistons and liners and the FWA crank, Coventry Climax further made over 1,300 of the 1220 cc FWE engine for the Lotus Elite. There were also some special 750 cc units made for Lotus to attack the Index of Performance at Le Mans. In 1957 a 744 cc unit, known as the FWC, was built using FWA pistons and a 45.2 mm stroke crankshaft; and a year later two 741 cc versions were developed from the tiny 633 cc marine derivative and called FWMAs (Featherweight Marine Automotive!)

One designer to respond to the *Motor Racing* appeal was Owen Maddock at Cooper. He dropped the FWA into the rear engined 'Bobtail', or T39, which had a chassis similar in concept to the Mark IX 500. Ivor Bueb gave the Bobtail its first run at the 1955 Easter Monday Goodwood, where he ran third to two 1500 cc Connaughts. The Bobtails dominated the 1100 cc class, when there was one, and often won in the 1500 cc or 2-litre class. In the early part of 1955, Chapman was using the old MG engine in the latest Lotus Mark IX, but by the middle of the season he started giving Cooper some competition with the Climax version.

For 1956 Colin Chapman produced the Series 1 Lotus Eleven. This came with a divided modified Ford swing axle front suspension, disc brakes, rack and pinion steering and, in 'Le Mans' guise (i.e. serious racing), a de Dion back end. Climax alternatives were either FWA or FWB, and the whole package was clothed in Frank Costin's prettiest aerodynamic bodywork. 150 of these Series 1 cars were produced between January 1956 and August 1957. To illustrate the impact they had on the 1100 cc class, the 32 entries for the 1957 August Bank Holiday Crystal Palace meeting were made up of 25 Lotus, 6 Coopers and the Halseylec: all, of course, were Climax FWA powered. Coopers may have been a bit short on numbers in 1956 and 1957, but drivers such as Ian Raby, Jim Russell, Moss, Salvadori, Brabham and Chris Bristow still gave the Lotus Eleven hordes a hard time.

The Series 1 Lotus, successful though it had been, did have some torsional rigidity problems, and that aged front axle gave difficulties with the steering geometry. In 1957 Lotus developed the Series 2 Eleven with wishbone front suspension, and more bracing in the scuttle area of the chassis. Just over a 100 of this model were produced before the end of 1958. The Series 2 dominated both 1100 cc and 1500 cc sports car racing in 1957, with no less than four Team Lotus cars running at Le Mans and coming first, second and fourth in the 1100 cc class, and winning the 750 cc class and the Index of Performance.

* * *

Chassis and suspension design the key to success

So this was the scenario that faced Eric Broadley when he sat down with a clean sheet of paper just before the end of 1957. The mission was clear: "Beat Lotus!" In fact, there was only one realistic strategy he could follow to achieve this. Everyone used the same engine, and, although at that time Eric fancied his chances as a racing driver, there were some fairly useful peddlers among all those Lotus Elevens too. No, the only way to go was to build a better handling car. There were two areas where he felt he could come up with something better: a chassis with a stiffer centre section, and a replacement for the heavy de Dion rear suspension.

Broadley's civil engineering background gave him a training in stress calculations in structures, and he put this to good use. The suspension layout of the new car was designed to feed cornering loads into the tubular chassis over a wide base, and for the chassis to absorb them progressively into the structure. The basis for the chassis multi-tube structure was to provide stiffness at the mounting points for the major components.

The front suspension loads were taken into 1-inch square tubes, via welded-in bushes. Loads were then taken out in various triangles to the the side and upper rails, and to a very stiff cruciform structure forming the forward bulkhead.

At the rear, fore-and-aft loads came into the chassis via the single 7/8-inch diameter radius arm. This was slightly curved on the prototype and early cars to clear the rear tyres. The load fed into the chassis via a mix of 1-inch square and round tubes, and 3/4-inch round tubes. The transverse loads from the rear suspension, which is described in detail later, were taken by the driveshaft and a lower link/wishbone in the shape of a roman numeral IV. The differential casing was rigidly mounted at top and bottom, and fed loads into the chassis with a short vertical member and others running forward into the transmission tunnel. The rear suspension Girling coil spring/damper unit was mounted at the top corner of the rear bulkhead, which was again 1-inch square tube, with transverse and forward bracing taken by 3/4-inch round tubes.

The centre section of the chassis was composed of the engine and scuttle bulkheads which were cross-coupled together in a number of planes. At one point the engine bulkhead cranked

A photograph of the recently restored chassis of the prototype. Note particularly how the rearmost 1-inch square front wishbone chassis tubes slopes forward rather than backwards, as on all later cars. Also contrast the straight across instrument panel hoop with the first illustration in Chapter 5: again a unique feature of the prototype.
(below) *Rear suspension showing the Elektron hub carrier with top wishbone made up of the driveshaft and trailing arm. The positive lower link/wishbone location is also clearly seen in this photograph.*

back, to accommodate the rear of the engine, and was braced in a seven tube mounting point. Because of the good rigidity between these bulkheads, the cockpit sides were relatively low without compromising the structure. The engine and gearbox mountings coincided with the rear lower front wishbone position and scuttle bulkhead respectively.

Sif-bronze welding was used for the 20 swg mild steel seamless tubing in the chassis, making for a light structure weighing in at only just over 60 lbs. Typical of the quality detail was Broadley's approach to the various square to round tube translations, which were scarf jointed rather than just simply butted.

The front suspension wishbones were of equal length on the prototype, and both were coupled to the ubiquitous Standard Triumph uprights and inner Metalastik bushes. Front suspension was not adjustable.

The rear suspension was rather special. The first design Eric came up with was a swing axle arrangement, similar to a Mercedes-Benz, and with a very low roll centre. The Broadley cousins got as far as having castings made up, but then Eric decided it would not work. After going back to the drawing board, the final system adopted used a top wishbone composed of the trailing arm and the driveshaft itself. This system avoided spline lock up, which had bedevilled some other independent rear suspension layouts, although Hardy Spicer were less than enthralled at the prospect of what had been designed for Morris Minor propshaft universal joints, taking racing suspension loads! They insisted that the driveshaft between the two joints had a tie-rod circlipped in, so as that the wheel would not immediately fall off if the shaft failed. (Colin Chapman had also used the trailing arm/driveshaft arrangement on the rear of the 12, 15 and early 16s but, in effect, as the lower link, with the strut itself locating the wheel.).

The roman IV lower link/wishbone mounting, used by Broadley, controlled steering toe movement with a very positive lateral location, and incorporated an eccentric in the hub carrier to adjust toe-in. The hub carrier itself was cast Elektron, made by Stones of Charlton, from a pattern carved by Eric. As subsequent events demonstrated, the system was a significant step forward in suspension thinking, and became standard wear on later 1960s Formula 1 cars.

Wheels and brakes

The 15-inch magnesium wheels were Cooper, as used on the 1957 T43 F2 cars. The inboard side was machined to a flat face at the rear for a flange mounted on the outboard drive shaft. At the front the integral taper roller bearings were left within the wheel hub to reduce unsprung weight compared to a separate hub flange. The prototype initially ran on Continental racing tyres, but later Dunlop R5s would be fitted. The steering rack was modified Morris Minor inside a fabricated casing, and was operated through a straight steering column, devoid of universal joints, for maximum positive feel. It had $1\frac{3}{4}$ turns, lock to lock.

Disc brakes had become pretty much indispensable on British sports racing cars, since Jaguar and Dunlop had pioneered them in 1952. It was somewhat surprising, therefore, that Broadley plumped for drum brakes. Presumably he decided that these would be at no particular disadvantage on a light car in British National sprint races for which the Lola was primarily intended. The brakes themselves, outboard at the front, inboard at the rear, were based on the Triumph TR2 units, but mounted on special alloy backplates at the front for improved cooling. The shoes were shod with Ferodo VG 95 linings and actuated by MG TF pistons onto 10-ins x $2\frac{1}{4}$-ins front, and 9-ins x $1\frac{3}{4}$-ins rear, Wellworthy Alfin drums. The prototype, and the early cars, had radial fins on the rear drums — but later these became circumferential.

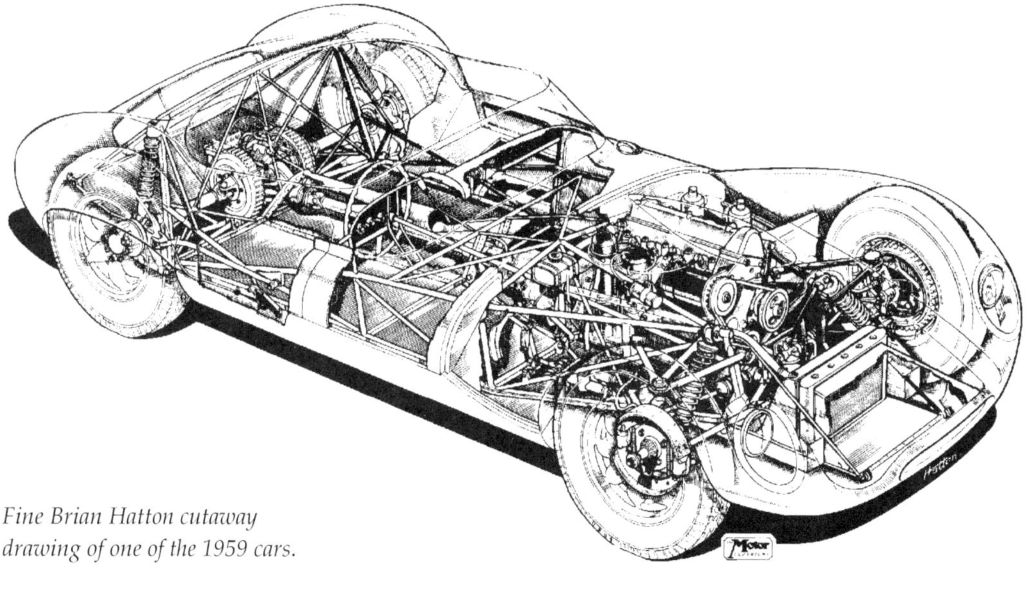

Fine Brian Hatton cutaway drawing of one of the 1959 cars.

Drive train

The Coventry Climax FWA initially used twin 1½-inch HD 4 SU carburettors, inhaling from a still air box, because the Broadleys could not afford to stump up the £80 plus for a pair of Webers. The gearbox used on the prototype came from an Austin A30/35 and used Lotus close-ratio gears. Later cars also used MGA Twin-cam and ZF gearboxes. The differential also came from an A30/35, and had a ratio of 4.55 to 1. On the prototype this fitted a fabricated steel casing, although this was replaced on later cars with a cast Elektron version, again from an Eric Broadley carved pattern.

Bodywork

The attractive aluminium bodywork, which brought rave reviews from the press of the day, happened by accident rather than design. Eric went to Maurice Gomm's small workshop at West Byfleet to discuss manufacture of a body. Gomm just happened to have an unfinished bonnet that he had made up for John Tojeiro, but did not look as though he would actually get paid for it. It was wheeled up and welded, but required very little modification to fit the Broadley's new car. Gomm and Eric Broadley sort of eyeballed the remaining bodywork to blend in with the lines of the front. So much for stylists and aerodynamicists! The two main body sections opened from hook type hinges at the extremities of the chassis, and the scuttle between the two centre bulkheads was fixed. The bodywork was mostly 20 swg, with the wheel arches and other minor items being 22 swg. The 5½ gallon aluminium fuel tank was placed on the nearside of the car, opposite the driver, in the interests of good weight distribution.

Not terribly good pictures (left and overpage) *but nonetheless historic.*

Just finished! The shiny aluminium prototype emerges from the lock-up behind 137 High Street, Bromley, in July 1958.

The dry weight of the car was about 830 lbs, some 35 lbs less than a Series 2 Lotus Eleven. Distribution was nearly 50/50 with the driver aboard, and basically was set up to understeer to Eric's liking. It had a 7ft 1-ins wheelbase and 4ft track.

The whole project required only just over eight months from conception to the first outing.

It had taken hours and hours of spare time dedication in the lock-up garage behind Broadley Bros Ltd (Outfitters) at 137 High Street, Bromley.

CHAPTER 3

1958: A Half-Season of Heartache and Promise

THE 'GREAT Exhibition' of 1851 had been celebrated by the construction of a large glass building, designed by Sir Joseph Paxton, to enclose the exhibits, and built on a site only eight miles from the centre of London. By 1927, motorcycle 'path racing' had come to the venue and, in 1936, work also started on a 2 mile paved road circuit for car and motorcycle racing. Sadly, also in 1936, a small fire got out of hand and Paxton's masterpiece in glass was effectively burnt to the ground. However, the racing circuit was completed and, after the Second World War, it was shortened to 1.39 miles when reopened again in 1953.

Crystal Palace: 5th July 1958

On Saturday 5th July 1958, the meeting was organised by the BRSCC on behalf of the track's owners: the London County Council (LCC). The weather was dry, but overcast. A relatively small crowd of some 6,000 turned up to watch a club meeting with production sports cars, Formula 3 500s, saloons, and double races for Formula 2 and unlimited sports cars. The feature 'Anerley Trophy' race for 1500 cc Formula 2 was dominated by Coopers, with Syd Jensen's Kiwi Equipe T45 leading Ivor Bueb's well driven Lotus 12 and the remaining Surbiton flock.

START 3.25 p.m. 20 LAPS	EVENT FOUR
For unsupercharged sports cars of unlimited engine capacity complying with Appendix 'C' to the International Sporting Code.	**The London Trophy Race** PART ONE

No.	*Entrant and Driver*		*Car*	c.c.
41	Brian Lister (Light Eng.) Ltd. (*Driver:* I. Bueb)		Lister-Jaguar	3781
42	J. C. Broadhead (*Driver:* R. Harris)		Jaguar D	3442
43	J. Sieff		Jaguar D	3442
44	Speedwell Garage (*Driver:* J. Bekaert)		H.W.M.-Jaguar	3442
45	P. Mould		Cooper-Jaguar	3442
46	N. Hillwood		Lister-Jaguar	3442
47	A. G. Whitehead		Aston Martin DB3S ~~or Lister-Jaguar~~	2922 ~~3442~~
48	L'Equipe Derbian (*Driver:* J. Dalton)		Aston Martin DB3S	2922
49	B. Halford		Lister-Jaguar	2986
~~50~~	~~Equipe Endeavour (*Driver:* Sir G. Baillie)~~		~~Lotus~~	~~1960~~ N.S.
~~51~~	~~L. Monteiro~~		~~Porsche 550RS~~	~~1498~~ N.S.
52	Weldangrind Ltd. (*Driver:* S. G. Young)		Parson-Maserati	1484
53	M. Trimble (*Driver:* D. S. Shale)		Lotus	1460
54	G. M. Jones		Lotus	1220
55	R. J. W. Utley	(Reserve)	Tojeiro	1098
56	E. H. Broadley	(Reserve)	Broadley	1098
~~57~~	~~M. B. McKee~~	~~(Reserve)~~	~~Elva~~	~~1098~~ N.S.

RESULT

1st 41 BUEB (LISTER-JAG) 2nd 49 HALFORD (LISTER-JAG)

3rd 47 WHITEHEAD (ASTON-MARTIN DB3S) 4th 54 JONES (LOTUS)

Winner's speed 76.94 m.p.h.

Fastest lap Car No. 41 at 79.43 m.p.h.

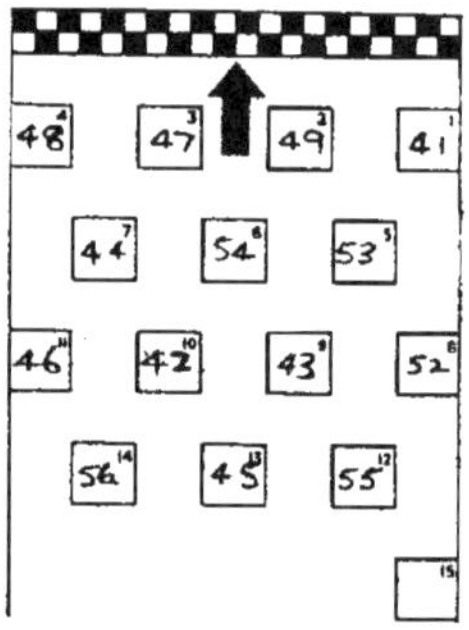

25

This page from the Crystal Palace programme shows second reserve for the 'Broadley', car No. 56, on its maiden outing.

The Palace was the Broadley cousins' local track, being only 5 miles away from Bromley, although neither had raced there before. For this first outing, the car was entered in 'The London Trophy' unlimited sports car race as a 'Broadley', not as a Lola. Two other small 1100 cc machines had entered and, together, the three FWA engined cars had been confined to be the reserves. However, fortunately, there were also three non-starters on the day, so the new Broadley got a place on the grid.

No 56: first race for the prototype in the London Trophy at Crystal Palace on 5th July 1958.

The new sports car was finished the previous evening in time for morning practice. Not having turned a wheel, Eric took things fairly gently all day, and was totally unaware that he did not have to run-in the Climax engine. The Broadley was slowest of the fourteen starters and found itself sharing the back row of the grid for Part 1 of the London Trophy with Peter Mould's Cooper-Jaguar and Richard Utley's Tojeiro. The little silver car finished, as it did again nearly two hours later in Part 2 of the race. Altogether an inauspicious debut, but the future potential was signalled by two lap times in the second race, only a little adrift of Bueb's race winning works Lister-Jaguar.

The First Win

Three weeks after Crystal Palace, the car had its second outing at Snetterton. This time it was running as a 'Lola', and with strong opposition in the 1100 cc Sports Car race. Keith Greene's Gilby Engineering Lotus Eleven won, and put in fastest lap at 89.01 mph. The Lola came second, in spite of a spin, in only its second meeting, and already gave indication that the old order was about to change. Early in the race, Broadley had also been led by one Peter Ashdown in his Team Lotus Eleven: Ashdown retired on Lap 3, but his path was due to cross Eric Broadley's again in a couple of months.

The first weekend in August was the traditional Bank Holiday, and a huge crowd of 54,000 assembled at Brands Hatch to watch the race meeting at the Kent circuit. The 'Rochester Trophy' event was for 1100 cc sports racing cars, and was set for two heats and a final. The single Lola had opposition from eighteen Lotus Elevens as well as other manufacturers. Heat 1 was won by Alan Stacey's Lotus Eleven. The Lola was in Heat 2 and, in practice, on the Saturday before the event, had broken the then unlimited sports car record with a lap of 59.4 secs.

Autosport reported the story of the race as follows:

> "Heat 2 brought the sensation of the day, which was provided by Eric Broadley's Lola. The little Coventry-Climax engined car was visibly quicker through the corners than anything else, and in practice it had been the first sports car to lap the Hatch in under a minute. Broadley immediately piled up a huge lead to win by 28 secs from John Brown's Elva, and equalled the class lap record of 60 secs in the process. . .
>
> In the Final, Brown's Elva shunted (the) Lola on the starting grid, which may have disconcerted Broadley. At all events he spun at the beginning of Bottom Straight during which he motored across the bows of several horrified competitors without disaster, and received the black flag for his pains. . ."

1958 Tourist Trophy at Goodwood. Peter Gammon goes outside the works Lotus Eleven of Alan Stacey and Keith Hall.

What John Bolster's report did not say was that Eric had made it up to fifth place halfway round that first lap, after being left in the starting skirmish.

Broadley entered the car again in the later 'Chequered Flag' 1500 cc race, and was forced to start from the back of the grid. In spite of this handicap, he passed twelve cars to finish fourth and put in fastest lap in the process.

The 23rd RAC Tourist Trophy... but first, an accident

Three weeks after the Brands Hatch exploits, Eric and the Lola were at Goodwood for the 31st BARC Members' Meeting. Again, the silver car had to start from the back of the grid. All was going well in the race, however, and by Lap 9 Broadley was pushing Keith Greene's Lotus for second place when they came up to lap a tail-ender at the double apex Madgwick Corner. Unfortunately, in the confusion, the back-marker spun and 600 DKJ (as the car had been registered) hit the bank and overturned. Eric was fortunate to escape with only concussion, and spent the night in hospital.

Peter Gammon takes a tight line at Woodcote in 600 DKJ.

The Tourist Trophy had been revived at Goodwood on 13th September after an interval of three years. In 1958 it was a round of the World Sports Car Championship, scheduled to last 4 hours. This was by far the most important event so far for the car and its constructors. It had been entered in the 1100 cc class, and Reg Tanner at Esso had offered £100 start money. Damage from the previous Goodwood outing included a bent chassis, suspension, and two wheels, as well as mangled bodywork. The two cousins worked virtually day and night for the whole period to repair the damage, and grafted a new front on to the chassis. They just about made it in time, with additional help from Eric's brother Richard, who made a hood at the family tailoring business to comply with international regulations.

Eric's co-driver for the race was Peter Gammon, who had initially made contact with the Broadleys when he had bought a GSM Delta road car from Graham. Gammon, who was rather appropriately a Guildford draper, had first made a name for himself driving a much modified MG TC. In 1954 he transferred the engine and gearbox to one of the first Lotus Mark VI, registered UPE 9. This car was fabulously successful taking fourteen wins, two seconds, and a third, from its seventeen starts.

This outstanding effort nevertheless did not lead to a close relationship with Colin Chapman when Gammon and ACBC fell out over an order for a new Lotus Mark VIII.

Reputedly, Gammon had ordered a new Mark VIII to replace the Mark VI and put down a £100 deposit. As construction got underway, he got to hear that ACBC's own 'identical' car would have lighter gauge tubes, thinner bodywork and one or two other rather special tweaks, so Gammon cancelled his order, and asked for his deposit back. Needless to say, Chapman was not about to comply with this request! Meanwhile, Gammon had previously had access to spares for the Lotus Mark VI model, and acted as a sort of agent for Lotus by selling these at circuits to other competitors. He got rid of the remainder of his stock and sent Chapman a cheque for the value, less £100. Chapman was not amused, and future relationships between the two men were non-existent. As a result, Gammon went with Cooper Bobtails in 1955 when the Climax engine came into small sports car racing. He had a few good drives with one of John Ogier's

Tojeiro-Jaguars in 1957, and was an ideal and experienced co-driver for the Lola.

Practice for the TT did not go well. Eric, weary from hardly any sleep for the previous two weeks, changed to first gear instead of third, with dire consequences for the gearbox. The Lola was dragged back to the Dennis lorry factory at Guildford, and a team of helpers set about taking the engine out. The gearbox was rebuilt and reinstalled, and the car returned to Sussex in time for 8 am scrutineering, although the carburettors and manifolds only just managed to get screwed back on as it waited in line for the officials.

Class opposition came from eight Lotuses, two of which were works cars, and three works Elvas, one of the latter, incidentally, driven by American, Carl Haas, who would play a significant part in Lola Cars' destiny in years to come. Yet again, fortune did not smile on the Bromley team. Early on, the car suffered a puncture from a cotter pin, when Gammon was driving, and the resulting sideways moment at the Chicane nearly tripped up Moss's Aston Martin too. Without any spare wheels left from the accident three weeks before, there was no alternative but to take the tyre off and repair the flat. This was carried out in about 15 minutes with the help of a spare inner tube scrounged from Dunlop. After that drama, the Lola put up the fastest lap in the class, and continued to run well until about 10 minutes before the 6 pm finish when the jack shaft driving the distributor and oil pump failed, and the car was eased over the line on the starter motor to finish sixteenth, and seventh in class. The 1100 cc class had been won by the Peter Ashdown/Gordon Jones works Lotus Eleven, from the Gilby Engineering car with Michael Taylor co-driving, with Keith Greene. The overall race winners were Stirling Moss and Tony Brooks leading an Aston Martin DBR1 one-two-three.

Enter Peter Ashdown

Peter Ashdown was born in 1934. At school, he ran a Matchless motorcycle, which was often used to travel to Brands Hatch "just to see what was going on." He signed on for three years' service in the RAF when he was eighteen, and learnt enough from a mechanics' training course to work on his own racing cars in later years. His first car was a Dellow. The Ford engine was tuned, and the car used in hill climbs, trials and driving tests of the day. He left the RAF in 1955 and began to work in the family coach company at Danbury, near Chelmsford. He also bought a Lotus IX in bits and pieces from Colin Chapman, as and when he could afford it, and providing Chapman had not sold the part he was expecting to collect to someone else first! By the time the FWA engine was eventually installed, and the car was up and running, most of the 1955 season had slipped away. Over the winter the car was updated by replacing the live rear axle by a de Dion system, and having Don Moore rework the Climax head.

Peter Ashdown: a man who played a major part in Lola's success over the next two years.

Theoretically, the Mark IX should have been beaten by the newer Series 1 Lotus Elevens in 1956. In fact, this was far from the case, as Ashdown recalled in a *Motor Sport* article some years ago:

"I had no difficulty with all the Elevens. One reason was that my 9 had drum brakes, and the Elevens had discs which tended to snatch."

This is an interesting aside, in view of Broadley's decision on this component, and may explain some of the immediate empathy Ashdown later had with the Lola Mark 1.

With the Mark IX, he had been so successful that he had won the John Coombs Trophy for "the best performance by a Lotus privateer", and Chapman invited him to join Alan Stacey and Keith Hall in the 1957 Lotus Eleven works B team (in fact, works supported owner/drivers). The three drivers had an excellent season with the cars and shared wins fairly evenly amongst each other.

Things looked good for the future, but early in 1958 he had a major accident at Rouen in the Team Lotus car in practice for the 'Coupe Delamare Debouttcville' meeting on 7th June. He broke his collarbone when the car spun and overturned on the fast downhill, twisting section, just after the pits. In fact, he was lucky this was his only injury, since the ambulance driver chose to head off on the verge of the circuit in the wrong direction, and was collected at some speed by Ron Flockhart's Coombs' entered Lotus 15. He was not fit to race again until the race meeting at Snetterton at the end of July, where Broadley was giving the Lola its second outing.

Several other starts resulted in only places until the TT class win got him noticed again, but at the end of the season the future with Lotus looked uncertain.

In October, Frank Nichols invited Ashdown to test his company's new Elva Mark IV at Brands Hatch, with a view to leading his works team in 1959.

As Ashdown recalled in *Motor Sport*:

"The car wasn't quite right but Frank was so enthusiastic that I felt we could possibly work together over the winter and put together a good package. I told Frank that I'd be in touch and, at the moment he and the Elva left the circuit, I was ready to drive for him. I decided to hang around and see out the rest of the day when Eric Broadley appeared with his car."

Meantime, Eric himself had been pondering on some words Nick Syrett, the BRSCC Secretary, had spoken at the August Brands Hatch meeting. Syrett suggested that Broadley might be better advised to put somebody else in the car to get the best out of it. Not surprisingly, this proposal got a frosty reaction from Eric, who had sweated blood over the car and had, at least up to that point, got a reasonable reputation as a driver.

But when Ashdown accepted Broadley's invitation to test the car, his performance made it clear to Eric that Syrett had a point. The two men immediately concluded a gentleman's agreement for Peter to drive for the team in 1959.

Ashdown again:

"I was convinced from the start. The car was superb on the track and, more, I liked the thought that had gone into it. Eric had designed it for strength and safety as well as speed, the wishbones were properly made and gusseted, for instance, and Eric's attitude was so different from Colin's (Chapman). We agreed there and then that I should drive for him and over the next few seasons I often had to shake down a new car in a race which was intended for a customer overseas. In that way I drove a lot of the Lola Mark 1s produced and often raced them without having previously seen them, and every one felt exactly the same, sweet and true."

* * *

A new Company: c/o KP Engineering, West Byfleet

The performance of the prototype in the latter half of 1958 was starting to attract a lot of interest for replicas for the next season. Then, when Eric 'went sick' for the two weeks before the Goodwood TT, his employers, Treasures, were again less than impressed and started to hint that their Contracts Supervision Manager might just be better employed by another company!

Furthermore, Peter Gammon had been so impressed by his drive at Goodwood that he was prepared to buy the prototype for £1,200. Additionally, soundings of the oil companies had been taken, and Shell had promised a further £1,000 for the kitty if any 1959 Lolas ran on their products. Then, finally, there was the Ashdown test, so that Eric decided it was time to draw his remaining pennies out of the bank and set up as a full time racing car constructor. Typically, he planned to build only four more cars for the 1959 season. He believed it was more important to do the job properly and stay close to his customers, rather than flood the market with cars whose quality he could not control to his complete satisfaction.

Clearly, the Bromley lock-up garage would not be suitable for building and operating a small team, so other premises had to be found in a hurry. The first home of the fledgling Lola Cars Ltd would be Maurice Gomm's shop at West Byfleet. Hence, if you had wanted to order your car for 1959 you would have written initially to KP Engineering, 3 Parvis Road, West Byfleet, Surrey.

A letter from America

One of those who did write (on 2nd October 1958) was an Allan Ross from Chicago. Among a number of interests, Ross was a restauranteur and amateur racing driver, who had made a name for himself driving an Elva during 1957 and 1958 at nearby Mid-West tracks. He had read the rave reviews in the English racing press about the Lola and enquired about the possibility of buying a car. By the following January, a deal was done to sell him one of the 1959 cars. He also made one further stimulating suggestion in his letter: if Lola were interested in exporting cars, he would be prepared to act as their distributor.

We shall return in Chapter 6 to Ross's beginnings in a market which would come to prove a major factor in the life of the Company over the next four decades.

Boxing Day Brands: a postscript

600 DKJ ran in one more race in 1958, and it was rather typical of the season. Peter Gammon, by now the owner, had entered it in the then traditional 26th December, post-Christmas race meeting at Brands Hatch. In the 1100 cc sports car event, Graham Hill had claimed pole position with a rather special Coventry Climax engined Lotus Mark VII. David Piper's Lotus Eleven led from the start, but he got a 10 second penalty for being "half way to Druids by the time the flag fell." Nevertheless, he was chased hard by Peter Ashdown having his last race for Team Lotus before switching to Lola. There was a coming together at Druids on lap 7, which let Hill up into the lead to win with Peter Gammon and the Lola a very close second, ahead of Piper, Ashdown, and the rest, including the three works Elvas. This result rather summed up the whole season for Lola: lots of potential, very nearly there... but not quite.

1959 would change all that.

CHAPTER 4

"Manufacturer of Competition Automobiles" The key 1959 Season

THE NEW Lola Cars Ltd planned to build four cars for the 1959 season for a determined attack on the 1100 cc sports car class. Although they were at first glance to look identical to the 1958 prototype, there were a number of significant changes for the first production run.

Subtle changes for the new cars

Perhaps most important were the refinements to the rear suspension. Colin Chapman had been indulging in a bit of gamesmanship and putting it about that the Lola rear suspension had a constant camber change due to the hinging of the bottom arms. To some extent this was true, and Eric Broadley recognised this by a redesign on the new cars. The lower wishbone was stiffened with a wider base, and made adjustable by having upper and lower inboard mounting points. The upper one gave a higher roll centre to give more warning of breakaway, and the lower one more wheel camber on bump or roll to increase cornering power but gave less warning of loss of adhesion. The other adjustment was at the outboard end, where an eccentric bush in the Elektron hub casing could be used to set negative camber by effectively varying the wishbone length so as not to alter the bump/camber ratio. The new cars ran at 1½ degrees.

The batch of 1959 cars under construction at Maurice Gomm's West Byfleet facilities.

Elsewhere, the front suspension was revised by moving the top wishbone anchorage point backwards a couple of inches to provide a wider angle to reduce flexing under braking. The associated 1-inch chassis tube was moved a little backwards of vertical, and the bushes sloped to improve anti-dive. The differential casing was now a Stones of Charlton-made, Lola designed, Elektron casting, rather than the fabricated steel unit that had been used on the 1958 car. Eric had carved and varnished the pattern for the casting himself.

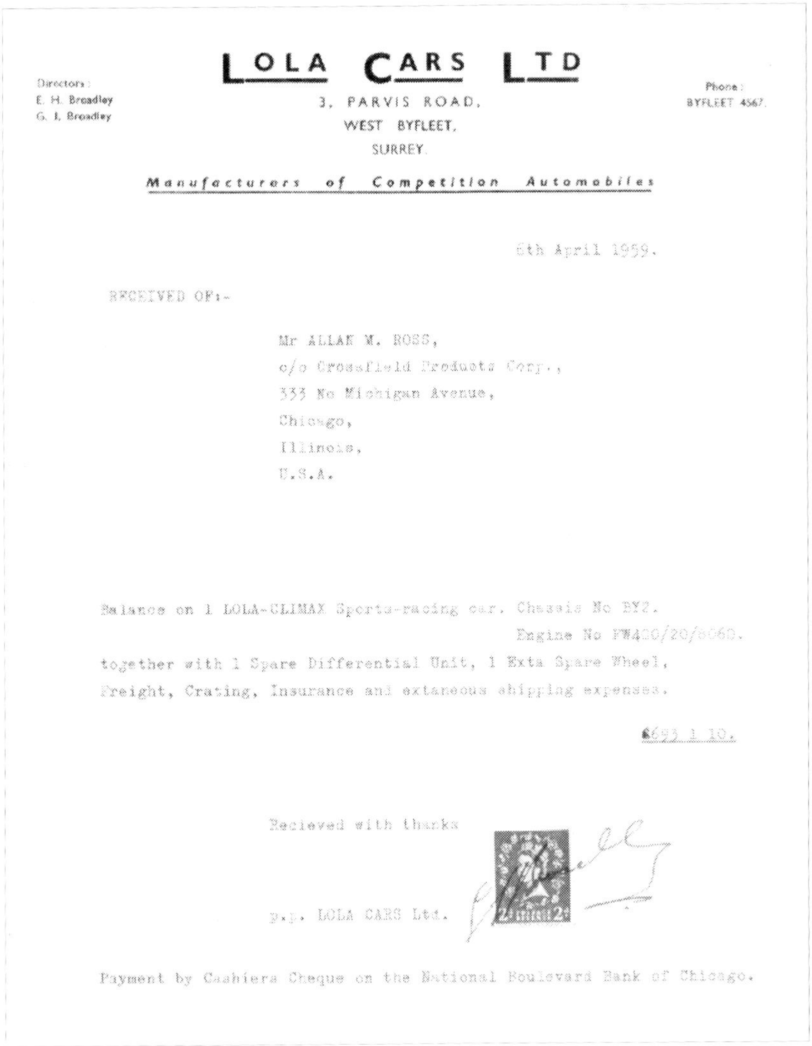

LOLA CARS LTD

Directors:
E. H. Broadley
G. J. Broadley

3, PARVIS ROAD,
WEST BYFLEET,
SURREY.

Phone:
BYFLEET 4567.

Manufacturers of Competition Automobiles

6th April 1959.

RECEIVED OF:-

Mr ALLAN M. ROSS,
c/o Crossfield Products Corp.,
333 No Michigan Avenue,
Chicago,
Illinois,
U.S.A.

Balance on 1 LOLA-CLIMAX Sports-racing car. Chassis No BY2.
Engine No FW400/20/6060.
together with 1 Spare Differential Unit, 1 Exta Spare Wheel,
Freight, Crating, Insurance and extaneous shipping expenses.

£695 1 10.

Recieved with thanks

p.p. LOLA CARS Ltd.

Payment by Cashiers Cheque on the National Boulevard Bank of Chicago.

The bodywork was changed around the cockpit area, and at the front and sides. The scuttle height was lowered 2½-ins with the result that the instruments needed to be housed in a raised binnacle. The latter required the windscreen to be moved forward slightly, and this was also thought to improve the aerodynamics. There was a minor adjustment to the radiator opening, which resulted in a lower bonnet line. The bonnet sides, where they met the undertray, now sloped downwards toward the wheel rather than the horizontal of the prototype. The other bodywork revision came later in the year with the completion of the Ross car. He had wanted a roll-over bar incorporated, and on his car there was a small hoop attached to the square tube running behind the seats to protrude (supposedly) above the driver's head. This provided pretty well zero protection, but it did create a streamlined driver's headrest to cover the roll-over bar and which characterised all subsequent Mark 1s.

In the engine department, the cars now breathed through 38 mm Webers rather than SUs.

A lighter flywheel had been tried, but discarded because it gave problems from standing starts. Gears had been a concern in 1958 and, for the new cars, Lola had designed special straight cut gears for the A30/35 gearbox cars with wide tooth base dimensions. For reliability reasons, Ross's car later became the first to be fitted with the MGA box.

It was also thought about time that the new company promoted its wares by having a proper bonnet badge made up. However, true to Lola tradition, you gave priority to things that improved performance; the first generation badges did not arrive until 1960!

The main opposition: Elva IV/V, and Lotus 17

Elva was another marque to have its beginnings in the 1172 Formula. They were initially designed by Frank Nichols himself, who owned the London Road Garage at Bexhill in East Sussex. The first car appeared in 1955, and in 1956 Elva were one of the first to use the new Climax FWA engine. The cars were never very successful in the UK against the Coopers and later Lotuses, but did achieve some sales success in the US by significantly undercutting rival firm's selling prices. For 1959, the company brought out the Mark IV, which was the work of Peter Nott, and largely an update of the Mark III, but with a new rear suspension, somewhat similar to that of the Lola. This also featured driveshafts as part of the top wishbone, but the beefy trailing arm was rigidly bolted to the hub carrier. The long, lower transverse links were not as widely spaced as the Lola. The Elva also relied on drum, rather than disc, brakes. Midway through the season, the company also brought out a lower, updated, Mark V version.

Nichols recognised quite early that his design ability was not in the same league as some of his contemporaries, and in

February he tried very hard to persuade Eric that he did not want to get involved in all this car building hassle, and that he ought to join Elva as their designer!

Lotus Engineering introduced the new 17 for 1959, which was reputed primarily to be the work of Len Terry. It featured a completely new strut-type front suspension, with the coil spring/damper unit almost vertical to minimise camber change on bump, and a wide spaced lower wishbone. At the rear, the Eleven's de Dion tube had gone, to be replaced by the strut and driveshaft trailing arm arrangement, similar to that used on the 12, 15 and 16. It was much smaller than the Eleven, having only a 6ft 10ins wheelbase and very narrow 3ft 6ins front track. Officially, it weighed in at only 750 lbs, although it must be said Colin Chapman's weights were often a bit lighter than the rest of the world's!

New Drivers, and a try out at Snetterton

The Lola 1959 driver line up for the UK was planned to be Peter Gammon in the prototype and, in the new cars, Peter Ashdown, Eric Broadley when available, and Bernard Cox, who was a relatively unknown former Elva driver. By now, the first of the customer cars had been sold to Taylor and Crawley Ltd for Clifford Taylor's son, Michael, to drive. Michael Taylor was another one of the rising star drivers who had come through the ranks of Lotus Eleven pilots in 1958. His car had been run by Innes Ireland Ltd, who looked after a fleet of six similar cars in the grounds of the Golden Acres Country Club at Elstead. Although he did get a run at a wet Le Mans with Ireland in one of the Team Lotus Elevens, by concentrating mainly on the less prestigious events, he was placed in all but one of his 27 races, and deservedly won the BARC Brooklands Memorial Trophy in his first season. The switch to Lola was a shrewd career move for this driver at the beginning of 1959, and the cash flow from the sale did the Lola Cars finances a power of good too.

The first meeting of the new season that Lola contested was the Spring Meeting held at Snetterton, where it faced mainly middleweight Lotus privateer opposition only. Three cars were entered for Ashdown, Cox and Michael Taylor, with the final one being so new that it was only completed on race morning at 7.30 am after several all night sessions. Bernard Cox could have wished for a better debut, since he damaged his car quite badly against a sand-filled oil drum at the Esses. The remaining two Lolas easily outpaced the field, with Ashdown leading Taylor home on the road in the 9 lap race. Peter, however, had been docked a minute for a jumped start—which, to this day, he swears he did not—but, even with this penalty added, he was still classified third. Rather appropriate, then, that Lola's first win of the year should go to their first private owner giving his car its race debut.

Goodwood – Easter Monday – 30th March

A huge crowd turned up to watch the usual BARC organised Easter Monday International meeting at Goodwood, which featured a 42 lap race for GP cars for the Glover Trophy. Lola had entered the first race on the programme for the Chichester Cup: a ten lapper for un-supercharged sports cars up to 1100 cc. The three Lola Cars Ltd entries for Ashdown, Eric Broadley and Gammon were joined by the Snetterton winning Taylor and Crawley car. Challenge came from twelve Lotuses, one of which was the Team Lotus 17 for either Alan Stacey or Graham Hill; two works and three private Elvas; and three one-offs, a Tojeiro, a Victoria, and a Moorland driven by the talented Chris Bristow.

Practice was on a dry track, although the morning was somewhat windy. Broadley non-started the fourth car which had not been repaired from Cox's accident. The three Lolas

Go! Ashdown, Gammon, and Taylor lead off the grid of the Chichester Cup at Goodwood on Easter Monday 1959 in front of a packed crowd.

BRITISH AUTOMOBILE RACING CLUB GOODWOOD

NEAR CHICHESTER SUSSEX

International Car Race Meeting

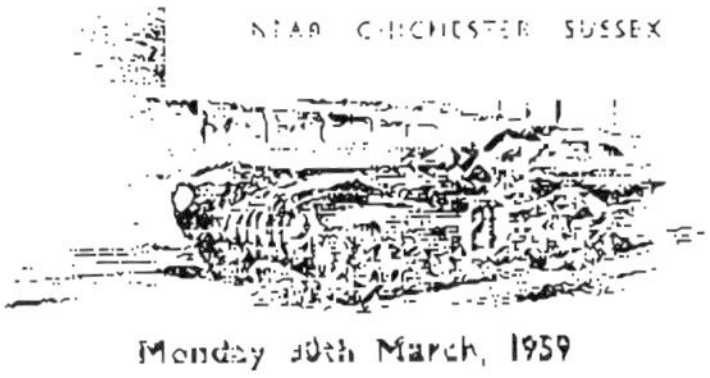

Monday 30th March, 1959

INFORMATION SHEET NO. 5

EVENT 1. THE CHICHESTER CUP 1.30 p.m.

(10 lap Scratch Race for Sports Cars up to 1100 c.c.)

Place	No.	Entrant and Driver	Car	Time M	Time S	Speed m.p.h.
1st	84	Lola Cars Ltd. (Driver: P. Ashdown)	Lola Climax	16	23.0	87.89
2nd	85	Lola Cars Ltd. (Driver: P. Gammon)	Lola Climax	16	28.0	
3rd	66	Taylor & Crawley Service Ltd. (Driver: M.J.C. Taylor)	Lola Climax	16	39.0	
4th	75	R.N. Prior	Lotus Climax	16	54.6	
5th	72	Cornwall Garages Ltd. (Driver: J. Campbell-Jones)	Lotus Climax	17	03.4	
6th	82	R.E.W. Utley (Driver: N.R. Hicks)	Lotus Climax	17	06.4	

Fastest Lap: Car No. 84 Time: 1 min. 35.6 sec. Speed: 90.38 m.p.h.

Non-Starters: 70, 71, 73, 76, 77, 83, 93.

Race Incidents: Car No. 74 (R.W. Waters) spun at Madgwick Corner on the first lap and retired.
Car No. 87 (J. Dickson) spun at St. Mary's Corner on the first lap and continued.
Car No. 90 (D. Hall) spun at St. Mary's Corner on the first lap and continued.
Car No. 78 (P. Riley) hit Chicane on the fifth lap and retired in the pits.
Car No. 66 (M.J.C. Taylor) spun at Woodcote Corner on the last lap, hit bank, continued and took third place to retire after the finishing line with damage to the rear of the car.

Special Item: Fastest lap constitutes new record for 1100 c.c. Sports Cars.

Goodwood Results

headed the practice times, with Peter Ashdown 0.4 secs clear of Gammon. Fourth was Alan Stacey's Lotus 17, 2.6 secs down on the pole time. Stacey complained of unsteadiness, particularly at Fordwater, which was just about a flat right hander in an Eleven. Hill took the car out for a three lap test, but also came off at

Fordwater each time. The car was withdrawn with 'suspension bothers', and fourth on the grid was taken by Scotsman Tom Dickson's Elva Racing Team Mark IV.

3rd April 1959 issue of Autosport *front cover acclaims the Goodwood achievement.*

The race was a Lola demonstration.

Martyn Watkins reported in *Autosport*:

> "This race was a Lola benefit: the marque's first appearance as a team could not have been more decisive, more impressive or more complete as a victory."

The three cars ran away from the opposition, such that Dick Prior's Lotus Eleven in fourth place was being dropped at 3 seconds a lap. Gammon led over the line for the first seven laps, although both Taylor and Ashdown had been ahead at various times. Then on lap 8, Ashdown took the lead for good, from Taylor, who made a last minute bid for glory at Woodcote but spun into the bank. Taylor restarted to keep the Lola 1-2-3, and still finish 15 seconds clear of Prior.

Ashdown's fastest lap was 1 min. 35.6 secs; 1.6 seconds below the old record, and only 0.4 seconds slower than Jean Behra's 1500 record with a Porsche RSK.

Same again at Aintree and Silverstone — and a spot of giant-killing at Oulton Park and Crystal Palace

Three weeks later, at Aintree, the same three Lolas repeated the Goodwood result, although Stacey's Lotus 17 had got on pole after the wet practice had masked the handling problems. Taylor led initially, followed by the Lotus, which broke its gearbox on lap 2, but finished behind Ashdown and Gammon at the chequered flag.

The 11th International Trophy was held at Silverstone on 2nd May. After the final Friday practice, Ashdown and Gammon were in the first two places, from Dickson's Elva, then Taylor and Eric Broadley out again in the repaired Snetterton car. Next up was Stacey's Lotus 17, with some of the 3/4-inch tubes now increased to 1-inch and different spring rates, but still in trouble. After the first lap Peter Ashdown was in his customary position from the other Peter, with Arundell's Lotus third. At 10 laps Taylor and Broadley had moved through the field to make it a Lola 1-2-3-4, but before the end Gammon retired with a broken oil pipe and Eric spun and put the back end into the bank at Chapel Curve. Ashdown won by 29 seconds from Taylor, with Dickson's Elva another 35 seconds away.

Between the Goodwood and Aintree outings, Ashdown and Taylor's repaired car were entered in the 1500 cc sports car race supporting the British Empire Trophy F2 race. Giving away over 30% engine output to the bigger cars, the "amazing Lolas"—as *Autosport* called them—finished second and fourth to Alan Stacey's Team Lotus 15.

Peter Ashdown ran at the BRSCC Whit-Monday meeting at Crystal Palace in the Anerley Trophy 1500 cc race, which was run in two heats and a final. The 1100 cc cars were in the second heat, and Ashdown had to be push-started because of a flat battery. He came first on the road, but was disqualified.

An aggrieved *Autosport* commented:

> "Although his win was by so great a margin, and although his performance had been such an impressive demonstration of Lola superiority, someone registered a protest, and the Stewards had no choice but to uphold it. Rules are rules and Ashdown was disqualified. Seldom has a race been won by such a convincing margin and the winner been struck out of the results. It was a demonstration of Lola high speed reliability and superiority and while the protest was fully justified, it nevertheless niggles a little. Further it kept Ashdown out of the final and ruined what would have been a first-class dice between the two Lolas and the two 1½-litre cars."

The reporter was clearly a paid-up member of the Ashdown admiration society!

Elsewhere, poor Bernard Cox's luck did not improve. At Mallory Park on Whit Monday, he was pushed off by Chris Threlfall's Tojeiro, and Peter Gammon cruised to a 20 secs win. Cox picked up a second to the Hon Edward Greenall's Lotus 15 at a Lancashire and Cheshire Car Club, Oulton Park, meeting in June, but was back in trouble at Mallory when the bonnet came off and again Gammon won. Gammon was by now driving the ex-Taylor 1959 car, which he had swapped for 600 DKJ since Taylor offered a deal when the latter considered that the prototype handled better than his own newer car. Gammon won again at the Leicestershire circuit in September in the 20 lap final for sports cars up to 1200 cc at the Nottingham Sports Car Club meeting.

Continental Excursions

By far the most important and ambitious race the works had tried so far was the Nürburgring 1000 Kms on 7th June. One car was entered for Peter Ashdown and Eric. Since they were looking at over 40 laps of the famous 14.18 mile Eifel Nordschleife, the car was going to need a little more thought on preparation than a Snetterton clubbie. They had been told that the 'Ring would require something like 64 gear changes a lap, and hence decided to fit the much beefier, but 30 lbs heavier, MGA 'box for the occasion. Suspension was also given some attention with Armstrong adjustable dampers, and a range of springs and packings were taken to try and eliminate bottoming at the Karussel.

Eric rounds the Karussel at the 1959 Nürburgring 1000 Kms.

Practice went fairly well, with the car taking chunks out of the class lap record, although there had to be an overnight engine change prior to the race itself.

Peter Ashdown recalled in *Motor Sport*:

"At the start of that race I'd put the Lola in eighth place on the grid. If you see photographs of the line up—we had a Le Mans-type start—you will see Ferraris, Maseratis, Astons, Jaguars and in place number eight you don't see anything. Our little car was completely dwarfed."

Sadly, Eric landed the car in a ditch just before half-distance when it began to rain, and handed the class win to the Campbell-Jones/Horridge Lotus Eleven. The Lola was undamaged, but he could not get it out by himself, and outside assistance was not allowed. While the car ran, however, it was miles and miles ahead of its class, and Ashdown had lopped a whole minute off the old class lap record. Moreover, he was only changing gear 14 times per lap: the old A30 box would probably have done the job.

On 12th July there was an F2 race at Rouen that was supported by the 4th Coupe Delamare Deboutteville, a 300 Kms race for up to 2-litre sports cars. Ashdown again ran the works car, but from early on the engine was making noises and he drove only just fast enough to finish fourth overall and beat the 1100 cc class opposition.

Two weeks later, Ashdown had perhaps his finest win, at Clermont Ferrand. Again, the two hour race, this time for sports and GT cars up to 3-litres, was in support of the F2 circus. The opposition was not perhaps World Sports Car Championship class, but one was: Jean Behra's 1500 Porsche RSK. Ashdown won the event outright by half a lap over the Stuttgart car.

Paddock scene at Rouen in July. Peter Ashdown (striped shirt) helps to pour 'l'essence'. Shell had put money into the Lola team in 1959, and no doubt the carefully posed oil cans were for the benefit of Shell-Mex and BP racing manager, Bryan Turle.

Back to Bromley, and other distractions

Eric Broadley's home was at Chelsfield, near Orpington. He had attempted at first to shuttle the more than 30 miles to and from West Byfleet but, 14 hours plus working days, mostly 7 days a week, meant that, as often as not, he slept at the Parvis Road works. As early as April, it was also becoming clear that Maurice Gomm's facilities were inadequate for race preparation and, additionally, the neighbours did not seem to appreciate the music of an FWA being run late at night.

The saviour of the situation was Rob Rushbrook. Rob had been involved in trials with MGs before the Second World War, and was also an RAC scrutineer. Post-war, he set up a garage business in Bromley with his partner. The car repair trade, in those austerity days, often needed a decent machine shop to keep tired old metal mobile, and 'Amos and Rushbrook' were just such premises, often taking in work for other local garages.

Eric Broadley first approached Rushbrook back in the early Ford Special days, to see if he could get the odd machining job done. In order to save money, he suggested that he be allowed to operate the lathes himself after the workshop closed. Although Eric was self taught, Rob was impressed with the quality of the young man's work. Shortly, Eric was back again to beg for a flaring tool to make up some brake pipes. Rushbrook became intrigued enough to see what his machine shop was contributing to nocturnally, and started going along to meetings as a helper for the Broadleys.

In the autumn of 1958, Eric told Rob he was going to set up Lola Cars to build several cars for 1959, and asked if he, Rob, would be interested in the project. Rushbrook at the time had said "no, unless it was at Bromley". In April 1959, Broadley called Rushbrook again to see if the earlier offer of space was still open. Rob's affirmative reply and question, "When do you want to start?" got a "Now!" reply from Eric. Rushbrook had some spare ground next to his workshop and the plan was to build an extension and turn the workshop over to Broadley. Eric's former building contacts proved invaluable, and the extension was completed in double quick time on the Stanley Road site. Legend has it that the yellow chassis colour of the later production Mark 1's just happened to be the same colour as the girders for the new extension.

By July 1959, Lola Cars Ltd were back at Bromley with a new Napier Road address.

It was also clear that something needed to be done about the business. Graham was still employed in the family tailor's shop, but doing his best to help out as often and wherever he could. Eric was trying to organise the move from West Byfleet, chase recalcitrant suppliers, support the racing activities, visit America to set up the Allan Ross agency, drive the odd race, plan for some more Mark 1s to be built, design a Formula Junior for next year and a few other things in his spare time!

It was clear to him, particularly after the 'offs' at Silverstone and the 'Ring, that he was not going to win a Driver's World Championship, and he decided that in future he would have to concentrate his skills mostly away from the tracks.

August Double at Brands Hatch

On August Bank Holiday Monday, the BRSCC staged a meeting at the Kent circuit in front of a record crowd. In the Wrotham Trophy Final for 1100 cc sports cars, Ashdown and Gammon again posted a 1-2 over Graham Hill's Lotus 17, with the former lapping in a sensational 57.8 secs. This time was comfortably quicker than any sports car had been round the one-and-a-quarter mile circuit at that time, and was only one fifth shy of Jack Brabham's F2 (1500 cc twin-cam FPF) Cooper!

Brabham was mightily impressed in his regular *News Chronicle* newspaper column, and wrote:

> "How do they (Lola/Ashdown) do it? It is obviously a combination of lightness and good road holding. The next time they're racing I'm going to give them a closer inspection."

His opportunity came three weeks later at the International Meeting held on 29th August.

Lotus had by this time given up on the strut front suspension on the 17 and fitted wishbones in an effort to solve the handling problems. Both Ashdown and Gammon were entered, having cleaned up the day before at Crystal Palace with yet another 1-2 and a 63.0 secs lap record. At Brands, the two Lolas were in the 15 lap Rochester Trophy against Graham Hill, driving the revised team Lotus car, Keith Greene's Lotus 17, Mike McKee's new Elva Mark V, and hordes of Lotus Elevens. The two Peters assumed complete command at the start, with Hill hanging on as best he could. After eight laps, Ashdown went off on the grass at Bottom Bend, but re-caught Hill for second place before the chequered flag, and equalled his Bank Holiday Monday record again in the process.

Stand here for the accident!

About this time the Triumph Herald had been introduced. Eric could not believe how anyone could design such a disastrous rear suspension geometry and, on the way to Silverstone one day, spent some considerable part of the journey outlining the shortcomings to Peter Gammon. After their race, the Lola had been loaded on the trailer for the trip back to Bromley and they were about to leave, when a saloon car race started: it included a Triumph Herald. They decided to watch, and stood on the outside of Woodcote. Eric said he reckoned if they moved about 20 yards to the right they would get an even better view of the accident. He was exactly right!

What a difference a year makes — The 1959 TT

The 24th RAC Tourist Trophy was held at Goodwood on 5th September, with Aston Martin this time needing an outright win to see off opposition from Ferrari and Porsche for the World Sports Car Championship. This they accomplished when Stirling Moss took over the Carroll Shelby/Jack Fairman DBR1 at 99 laps after his lead car, shared with Roy Salvadori, famously caught fire in a pit stop. The Aston completed 224 laps in the 6 Hour race, to win by a lap from the Bonnier/von Trips 1.6-litre Porsche RSK, and the Brooks/Hill/Allison Ferrari Testa Rossa.

Three Lola Cars Ltd Mark 1s were entered in the 1100 cc class for Peter Gammon/Dick Prior, Bernard Cox/Colin Escott and Peter Ashdown/Allan Ross (the latter over on a visit from the US). Up against them were three Elva Racing Team cars, two Team Lotus 17s (for Keith Greene/Tony Marsh and Innes Ireland/Jay Chamberlain), and two private Lotus Elevens (for John Campbell-Jones/John Horridge and Peter Arundell/Jack Westcott).

*1959 TT.
Peter Ashdown on the line at Madgwick in this classic George Phillips photograph.*

The programme of the day noted:

"Throughout the whole of this season the Lola cars have proved almost unbeatable and this comparatively new design has displayed road holding which is probably unequalled by any other sports car design at the present time."

Predictably, Peter Ashdown was quickest in practice with a time of 1 minute 36.8 secs, and he lined up for the Le Mans start in

*Lola Cars' 3 entries won the team prize at the TT.
No 34 is the 1958 prototype, driven by Dick Prior and Gammon, out-accelerating the McKee-Brierley Elva away from the chicance.*

sixteenth place between Scarfiotti's 2-litre Ferrari and Hans Herrmann's works Porsche. Nearest class challengers were the two Lotuses of Arundell and Ireland, both having lapped in 1 minute 37.4 secs. The other two Lolas were next up in the class.

In the race itself, Innes Ireland held an initial class lead but retired with a broken chassis and the Lolas dominated the class. The Ashdown/Ross car finished sixth overall, behind the two remaining Astons and Testa Rossas and, with the other Lolas placed eighth and tenth, Lola Cars Ltd collected the team award donated by The Society of Motor Manufacturers and Traders.

A helping hand for Bill

The BARC ran an annual contest for the Brooklands Memorial Trophy for the most points scored by a Member during the season. Coming to the last meeting of the season at Goodwood on 26th September, Triumph TR3 driver Bill de Selincourt faced opposition for the Trophy from Paddy Gaston and Tony Maggs. In order to build up a few extra points, he borrowed Peter Ashdown's car for the first 10 lap scratch sports car race. Eric himself was on hand to offer a few tips. Unfortunately, the advice did not run to starting, however, as de Selincourt made a poor getaway and trailed Chris Steele's Lotus Eleven in the early stages. However, he quickened up as he got familiar with the car and overtook at half distance for a reasonably comfortable 3 second win.

It was all worth it; and de Selincourt landed the Trophy for 1959... and he put an order in for a new 1960 Mark 1. Equally impressed was Chris Steele, who promptly sold his Lotus and bought the ex-Ashdown car he had just been beaten by!

* * *

A Christmas present for Alan Rees

Brands Hatch hosted the usual Boxing Day meeting to help disperse the Christmas pudding. By now, Peter Gammon had retired from motor racing when the apparition of the late Mike Hawthorne appeared to him through a medium at a spiritualist meeting and advised him to stop. Alan Rees had purchased the ex-Taylor/Gammon 1959 car, and was giving it its first outing in new hands. Local Bromley garage owner Chris Steele was also having his first Lola race in the ex-works Ashdown car.

The Christmas Trophy was for unlimited capacity sports cars, and included an award for the first 1100 cc car. Starting from the back of the grid, Rees simply drove through the field to hit the front on lap 4 passing, among others, John Wolfe in a Jaguar C Type. Wolfe was also to be impressed enough to order a Mark 1 for the new year.

The *Autosport* race reporter commented:

"After all, what's a few extra litres or so when the other chap's got a Lola!"

Alan Rees at the wheel of his Lola during 1960.

The First Racing Car Show

With the growth of the British motor racing car industry in the 1950s, it was surprising that it took so long to recognise that a Motor Show devoted to competition machinery would be popular. Ian Smith of the BRSCC set about organising something, and the first 'Racing Car Show' was opened at the Old Horticultural Hall, near Victoria, on 2nd January 1960.

Lola Cars had their new Formula Junior squeezed on Stand No 21, but the works Mark 1, as BRSCC Sports Car Champion, was selected to be on display in pride of place alongside the World Sports Car Championship winning Aston Martin DBR1.

* * *

Fine overhead late season shot of the engine bay of the works car.
Note the space for the filler cap of the second petrol tank that had been used on long distance races.

Shell capitalised on their 1959 investment in Lola Cars in their annual Shell Successes *booklet.*

(*Above*) Empire Trophy meeting, Oulton Park. **Parker** (Cooper, No. 12), who ultimately won the 500 c.c. race, fights a duel. (*Below*) Aintree "200" meeting, 1100 c.c. sports car event. Partly obscured by the opposition is **Ashdown** (Lola), who fought through to win.

Photos: "Autocar" and "Autosport"

28th March. EELMOOR PLAIN

RACE 4, 750 c.c. RACING CARS
1st P. Emery — Emery-Velo
RACE 7, COMBINED BERKELEYS & 250's
1st P. Emery — Emery-Velo

30th March. GOODWOOD

10 LAP SCRATCH RACE FOR SPORTS CARS UP TO 1,100 c.c.
1st P. Ashdown — Lola
2nd P. Gammon — Lola
4th R. Prior — Lola

30th March. BRANDS HATCH

"WORLD SPORTS" TROPHY FINAL (FORMULA 3)
1st D. Parker — Cooper
GRAND TOURING CAR RACE — CLASS B over 1,600 c.c.
1st J. Sears — Austin-Healey
CLASS E SPORTS CARS
1st I. Walker — Willment
EASTER TROPHY FINAL (15 LAPS)
1st I. Walker — Willment

5th April. BLANDFORD HILL CLIMB

750 TROPHY
J. French — Simplicity
VINTAGE CARS
J. French — Austin
750 FORMULA CARS
1st D. Hitches — Austin

WINNERS HANDICAP RACE
1st G. Williamson — Austin-Healey Sprite

10th May. MONACO GRAND PRIX

2nd C. A. S. Brooks — Ferrari

10th May. SNETTERTON

SPORTS CARS UP TO 1,500 c.c.
1st P. Ashdown — Lola
2nd B. Cox — Lola

18th May. CRYSTAL PALACE

TOURING CARS
1st G. Williamson — Alexander-Austin A.40

18th May. MALLORY PARK

SPORTS CARS UP TO 1,200 c.c. (HEAT 2)
1st P. Gammon — Lola
SPORTS CARS UP TO 1,200 c.c. — FINAL
1st P. Gammon — Lola
FORMULA 2
1st T. Parnell — Cooper

30th May. RUFFORTH

SERIES PRODUCTION SPORTS CARS AND GRAND TOURING CARS UP TO 1,600 c.c.
1st M. Wayne — Elva
SCRATCH RACE FOR RACING AND SPORTS RACING CARS
1st T. Taylor — Cooper F.2
3rd T. Parnell — Cooper F.2

Lola, the *dernier cri* in flyweight sports-racing cars. Sample pictured here is driven by **Peter Ashdown** at Silverstone in the B.R.D.C.'s International Trophy meeting. **Ashdown** dominated his race throughout, challenged only by another Lola pilot, and won over a 25-lap distance at 93.60 m.p.h. average. He uses Shell in sump and tank.

Photo: "The Motor"

Lola versus Lola, but both are on Shell X-100. Leader in this two-of-a-kind duel at the Crystal Palace is Peter Gammon, who won the 1100 c.c. sports car final. Pursuer is Peter Ashdown, who finished second.

Photo: "Autosport"

26th September. OULTON PARK

"AUTOSPORT" CHAMPIONSHIP
CLASS A
1st F. Gerard — Alexander-Turner
CLASSES B & C
2nd C. Meek — Elva Courier
CLASS D
1st J. Stoop — Frazer-Nash

10th October. 1959 "AUTOSPORT" FINALS, SNETTERTON

FORMULA 2
1st C. Bristow — Cooper-Borgward
3rd J. Campbell-Jones — Cooper
"AUTOSPORT" CHAMPIONSHIP FINALS
GENERAL CLASSIFICATION
2nd J. Stoop — Frazer-Nash
CLASS A
1st F. Gerard — Alexander-Turner
CLASS D
1st J. Stoop — Frazer-Nash
"AUTOSPORT" CHAMPIONSHIP — AGGREGATE RESULTS
1st { F. Gerard — Alexander-Turner
J. Stoop — Frazer-Nash }

CHAPTER 5

A Busy First Year at Bromley

WORK ON the Napier Road site was sufficiently progressed by the turn of the year for ambitious plans to be made for significantly increasing output of Mark 1s to meet the demand both in the UK and overseas for the 1960 season. Additionally, Formula Junior racing was catching on rapidly, and Lola had designed a car based on the sports car to compete in this new market. In fact, such was the reputation of the Mark 1, that twelve orders had been placed for the new single seater by January 1960.

Expansion plans

To cope with this expected level of production, Rob Rushbrook and the Broadleys began to search out suppliers and recruit staff.

Les Weller and Mike Hyatt joined machinist Paul Wilmshurst, —and usually a couple of mechanics from South Africa and Australia who were gaining experience, and also did not attract National Insurance contributions!

The list of some of the suppliers ran:

Chassis frames	Arch Motors, Tottenham
Wooden patterns	London Pattern Co
Alloy castings	Stones of Charlton

Seats	Connolly Bros, Leyton
Close ratio Gears	Mike Hewland, Maidenhead
Exhausts	Derringtons, Kingston
Aluminium panelling & petrol tanks	Shapecraft, Surbiton
Fibreglass bodies	Specialised Mouldings, Upper Norwood (*whose interface with Lola Cars is addressed further on*)

Skinned chassis of a 1960 car outside Napier Road.

To gather up all these parts, together with the Coventry Climax engines and Dunlop tyres, was Mike Hyatt's responsibility. The works transport was a hard-pressed old Ford 7 cwt Estate, registered 293 CPE, which, as often as not, was made to tow

Rob's trailer as well. It usually sported part-worn racing tyres, courtesy of the Fort Dunlop tyre fitters, which could be had for a £1 a time. On one occasion, Hyatt complained to Eric that the brakes needed an overhaul, to which he got the classic reply, "I pay you to go, not stop!"

Tuning for the works Coventry Climax engines was carried out by the legendary Bill Basson. Bill was ex-RAF, and had lost a leg and a hand during the war. He was Alan Stacey's mechanic initially, and Essex men Stacey and Ashdown were old friends, so Basson also started tuning Ashdown's engines back in his Lotus days too. His missing hand had been replaced by a hook, so it was a tremendous triumph over physical adversity to be regarded so highly as an expert engine tuner.

Specialised Mouldings Ltd

Historians argue about when man discovered plastic, but first claim went to J W Hyatt (no relation to Mike) in 1868, who found that cellulose and camphor, moulded under pressure, could be made to give a hard, attractive material suitable for detachable shirt collars and billiard balls. The first proper plastic patent was attributed to Dr Leo Baekeland in 1907, who made a wholly synthetic material from phenol and formaldhyde.

Polyester resins were developed in the 1920s and 30s and, by the Second World War, similar unsaturated polyesters were dissolved in styrene monomer to provide laminating resins. A decade later, such materials were being commercially polymerised at room temperatures with glassfibre lamination to form, for example, reinforced boat hulls.

Peter Jackson worked for a producer of plastic motorcycle side cars, but believed that there was potential for growth in other markets. During 1959 he branched out on his own to make a children's peddle car version of the Vanwall. Sadly, this venture lasted only four months, but it did bring him to the attention of Eric Broadley. Peter was aware that glass-reinforced plastic could, potentially, be used for full-sized racing cars, and that it could offer an improved strength/weight ratio over the then traditional aluminium. It was also easier to fabricate repeat orders, once a mould had been made, and could be simpler to repair.

Later in 1959, Jackson formed Specialised Mouldings Ltd, with his brother, after meeting with Eric. The two of them set about developing techniques together with the objective of supplying fibreglass bodies for Lola's Mark 1 production in 1960. As it happened, Eric was also taken with the mini-Vanwall peddle car too, and reckoned it was just the shape for his new front-engined Formula Junior, although in practice the bodies for these cars were actually made in aluminium.

Specialised Mouldings went from strength to strength on its involvement with the winning Mark 1 and, by the end of the year, were making bodies for Cooper and the new 1961 Lola Mark 3 Formula Junior. Within a further year, the company was also making bodies for Brabham and Lotus.

More evolutionary changes for the 1960 cars

Quite a lot of detailed changes were made for the latest cars. Most obvious, externally, was that the upper bodywork was now fibreglass rather than aluminium. The rear body included the headrest design first seen on Ross' 1959 car, and also the new bonnet extended rearwards to replace the separate scuttle section of the earlier cars. The lower edge of the bonnet became horizontal with the undertray again, as it had been on the original prototype. Additionally, the perspex-covered headlamps were moved forward to a lower position. The first of the new fibreglass bodies arrived at Bromley in late October 1959.

Under the skin there were several alterations too. Most major was that the engine was moved back 3-inches to improve weight distribution and counteract some of the understeer of the first cars. This relocation required some surgery to the chassis, and the 'seven point' tube junction was also moved backwards toward the dashboard to make the extra space. The brake and clutch master cylinders were elevated above the pedals rather than being at floor level. At the rear, the inboard lower wishbones were now non-adjustable, and the curved rear radius arms were replaced by straight ones, necessitating a modification to the hub carrier. Mechanically the cars were little changed except that some cars had ZF gearboxes fitted instead of the A35-derived or MGA units.

Supply of the old Cooper wheels was becoming a problem too and, additionally, a new 13-inch wheel was required for the new Formula Junior. Lola therefore decided to have patterns made and cast their own. The design was very similar to the former wheel, but a little smoother and without the holes near the hub.

A welcome Swiss customer

As with all small companies, the first stage of growth brought cashflow problems, but a saviour came in the shape of a wealthy Swiss called Charles Vögele. He wanted to do the major international meetings with Ashdown as co-driver. Lola were not overly keen on this arrangement as Vögele was not particularly rapid, and also they had limited control over car preparation, but the money came in handy. Hence, one of the first of the new 1960 cars was sold to Vögele and prepared for long-distance racing with an MGA gearbox and 15 gallon petrol tankage. The CSI had also brought into force their 'Appendix C' regulations at the start of the year. These required a 25 cm high windscreen, 12 cm ground clearance, and defined luggage space as well as other restrictions.

Vögele and Ashdown tested the new car at Goodwood over the winter before the car was shipped out to Sebring for the 12 hour race on the old Florida airfield circuit in March. In the Amoco sponsored race, Olivier Gendebien and Hans Herrmann scored a surprise victory in a Jo Bonnier-entered Porsche, when the faster Maseratis and Ferraris failed. The Lola came in 17th after being delayed by 40 minutes when the bonnet came adrift. Nevertheless, this was still good enough to win the 1100 cc sports car class from the American driven Costley/Tweedale (another later Lola convert)/Harrison Elva Mk IV.

Happy, if cold, group, winter testing Charles Vögele's new car at Goodwood. Peter Ashdown on the left, unknown, mechanic Robert Schittli and proud new owner.

First outing for the Swiss car at Sebring. Vögele/Ashdown line up ahead of the Scarlatti/Baron Serena (of Scuderia Serenissima), and Publiker/McCarthy Ferraris.

The works car shows off the new bonnet badge on 2nd April 1960 at Oulton Park. The hard-worked company Ford estate, 293 CPE, can just be seen to the right hand side of the picture.

New cars, new drivers

Internationally, sports car racing was changing from the golden years of the 1950s. The new Appendix C regulations had encouraged GT cars to enter major events, and 1960 was very much a mixture of these two approaches to long distance racing. In British Club racing, however, this distinction had yet to take hold, and 1960 still catered widely for sports racing cars.

Alan Rees romped home in the 1100 cc race at the opening Snetterton meeting on 26th March, setting fastest lap at over 90 mph. The following weekend, at the BARC Oulton Park meeting, Rees came home in front of Ashdown, who was giving the 1960 works car a first run with new ultra-lightweight bodywork which was so thin sparks could be seen through the bonnet when the engine was started! On Easter Monday, Rees and Steele climbed all over David Piper's Lotus 15 in the two part sports car race at the BRSCC Brands Hatch meeting, but had to concede defeat to the bigger car. David Hitches also gave his new car its debut in this race, and picked up second on aggregate when the other two Lolas struck problems. On the same day Bill de Selincourt debuted his new car at Goodwood, and Scotsman David Bertram had a successful first foray at the northern Thornaby-on-Tees circuit.

The Aintree International meeting for F2 cars on 31st April included a race for sports cars up to 1100 cc. Alan Rees won again from Chris Steele, with Keith Greene's Gilby Climax overtaking brewing family man the Hon Edward Greenall's new Lola within sight of the finish. By now six of the new 1960 models had been sold in the UK.

May was a busy month. At Mallory Park on 8th May the same three Lolas gave Brian Naylor's 2$^1/_2$-litre Cooper-Maserati a hard time in the unlimited sports car race, but had to settle for second, third, and fourth in the same order as at Aintree. Rees had a full weekend, with a third overall and class win the previous day at Oulton Park.

The International Trophy meeting at Silverstone on 14th May drew 80,000 spectators, although they were saddened by the untimely death of Harry Schell who had been killed in Friday's practice. Last race on the programme was for sports cars up to 1500 cc. Peter Ashdown was out again in the works 1960 car with its ultra-thin yellow fibreglass bodywork. Practice again

Ashdown demolished the opposition in the 1500 cc sports car race at the 1960 International Trophy meeting.

proved the Lola's superiority, with Ashdown lapping 4 secs quicker than Rees. On race day Ashdown broke the class lap record to leave it at 1 min. 48.4 secs, and win by 44 seconds from Rees, with the 1½-litre Lotuses of Geoff Breakell and Chris Kerrison next up. As it was becoming something of a trend, Kerrison was to order a Mark 1 too. The race also saw the demise of Laurie Keens' new Mark 1, which burnt out after an accident with Brierley's Elva and Greene's Gilby.

Nordschleife Stormtroopers

Lola were out in numbers for the 1000 Kms race at the Nürburgring on Sunday 22nd May. De Selincourt had recruited Chris Lawrence as co-driver, and they had found out the hard way at a previous Goodwood BARC Members meeting that their new Webers needed a bit of tuning; hopefully all would be right for the race. David Hitches had Bob Hicks to share driving duties, while Peter Ashdown was back in Vögele's car, and the Bertram brothers had entered their new car for its first taste of European racing.

Thursday it rained, as only it can in the Eifel mountains, and for Friday practice things were still wet. Quickest of the Lolas was the Hitches/Hicks car, with Vögele/Ashdown over 5 seconds adrift. Unfortunately, Douglas Bertram got something in his eye, so E H Broadley got drafted in as reserve driver with a promise that he would not visit the same ditch as he had in 1959.

Race day was still misty and drizzly, and was just made for another Stirling Moss epic victory driving the Camoradi Maserati Tipo 60 'Birdcage', together with Dan Gurney. On the whole, the Lolas ran well and dominated the 851 to 1150 cc sports car class. The Swiss car came through to win, from de Selincourt/Lawrence, who had some petrol filter problems, and early class leader Hicks/Hitches, who ran latterly without a clutch.

On the way back from Germany, all the Lola runners ran at the GP de Spa for mixed sports and GT cars. Unfortunately, the yellow works car was badly damaged when Bob Hicks was the innocent party in a major accident, after Jimmy Blumer's Cooper bounced off the bank at Burneville into his path.

Class win for Vögele and Ashdown at the 1960 Nürburgring 1000 Kms.

GrahamWhitehead, seen here behind the works car, had a one-off drive at the 4th GP of Spa on 29th May 1960, but did not really like small cars.

Adventure at Le Mans

The Automobile Club de l'Ouest (ACO) was founded in 1905, and organised the first national grand prix for the ACF following the earlier Gordon Bennett Cup races. On two sweltering days in June 1906, the ACO ran two races, each consisting of 6 laps of a 632 mile course. After the First World War, in 1919, the ACO proposed a new circuit of 10.8 miles near the town. A meeting was held a year later and, in 1921, the Le Mans permanent circuit staged the first Grand Prix de l'ACF after the war. In October of 1922 it was proposed that the circuit should put on a new form of motoring event based on endurance rather than speed. The first Grand Prix d'Endurance de 24 Heures took

24 HEURES INTERNATIONALES DU MANS

DEMANDE DE PARTICIPATION (4)

Conformément aux articles 1 § 2, 5 § 2 et 42 § 1, la demande de participation ne deviendra engagement qu'après décision des organisateurs, décision qui sera notifiée aux concurrents au plus tard le 10 mars 1960. Toutefois, sont invités d'office les concurrents Constructeurs qualifiés pour la finale de la XXVIe Coupe biennale (1959-1960) article 4 § 3.

Je soussigné Eric Broadley
adresse (complète) LOLA CARS LTD Téléphone : Ravensbourne 37 45
désire participer aux : Napier Road Bromley Kent

24 Heures du Mans 1960

(XXVIIIe Grand Prix d'Endurance)

avec le véhicule (1) SPORT conforme à l'annexe C du Code Sportif International
(1) ~~GRAND TOURISME conforme à l'annexe J du Code Sportif International~~
en vigueur au 1er janvier 1960 et dont ci-dessous les caractéristiques (2) (3)

a) Marque LOLA-Climax Type du véhicule Mark II
Puissance fiscale 5,5 Année de création 1960
b) Appellation du catalogue
c) Nombre de cylindres du moteur 4
d) Marque du moteur Coventry-Climax
e) Cylindrée avec mention de l'alésage et de la course 1098 cm3 71,37 x 66,6
f) Carburateur et sa marque Weber 38DCO3 Capacité réservoir carburant 79 lt
g) Dispositif de suralimentation pompe eletrique SU
h) Genre d'allumage et sa marque LUCAS bobine 12 V
i) Boîte de vitesses (genre et marque) MG 4 vitesse
j) Freins (genre et indication du genre du servo-freins utilisé) Lockheed 2 circuit freins tambour
k) Amortisseurs (marque et nature du fonctionnement) Armstrong hydraulique
l) Roues magnesium Marques LOLA Indépendantes ou non AV. et AR.
m) Dimensions des pneumatiques 15x450/15x500 (marque et genre) Dunlop Racing
n) Batteries d'accumulateurs (genre et marque) Lucas 12 V
o) Dispositifs d'éclairage avec leurs marques Lucas 12 V
p) Genre du pare-brise et matière du pare-brise Perspex
q) Noms des conducteurs, titulaires ou suppléants (par ordre alphabétique) Peter Ashdown (Angleterre) Charles Vögele (Suisse), Eric Broadley (Angleterre)

(1) Rayer la mention inutile.
(2) Établir un bulletin par voiture.
(3) Il ne sera tenu compte que des demandes de participation spécifiant la marque, le type et la cylindrée de la voiture.
(4) Joindre à la demande la liste des conducteurs envisagés et leurs palmarès.

Le Mans entry form.

Le Mans. Scrutineer gets down to business to check out passenger seat dimensions.

place on 26th/27th May 1923, with Lagache and Léonard's Chenard & Walcker the best of the remarkably high number of thirty finishers. The circuit was shortened slightly in 1929 and 1932, and modified again after the tragic accident in 1955. Nevertheless, the 8.36 mile circuit remained a superb endurance test of both man and machine.

In order to make an entry more acceptable to the ACO, the request for participation on the 25th/26th June 1960 came from

Le Mans. Love that hat Peter! Ashdown, Vögele and reserve driver Dr Honeggar before the start.

Le Mans. Taking things easy in the early stages.

Lola Cars Ltd rather than Vögele. The red and white car was entered as a 'Mark 2', since at this time Bromley considered the updated sports car as a second derivative, and this became a little confusing afterwards since this label also retrospectively got applied to the 1960 front-engined Formula Junior. Ashdown and Vögele were nominated as the drivers, with Dr Walter Honeggar of Berne the reserve.

Le Mans. Fuel stop at 6.40 pm.

The car was issued with start number '45' and entered in the sports car class between 1001 and 1300 cc, where the only opposition came from the 1147 cc Conrero Alfa Romeo of De Leonibus and Consten. The little Anglo/Swiss team based itself at the Hotel Modern, having travelled to France via Boulogne. Scrutineering was at 8 am on the Tuesday before the event. The Le Mans multi-post scrutineering system, in those days still at Les Jacobins, was notorious for its fastidiousness and inevitably the Lola had a couple of near squeaks. The ground clearance block was only traversed by the expedience of lifting the car up on the shock absorbers first, and the carrosserie examiner was not at all happy about the fact that the side number was on the rear wheel arch and not on the door!

Ashdown took the first stint on race day, and completed 29 laps in 2 hours and 38 minutes before a 4 minute hand-over to Vögele. The car was running 37th of the 55 starters and just ahead of the Conrero Alfa. Vögele then took 2 hours 56 minutes for his 30 laps: the stop at 9.38 pm requiring a 10 minute pause before Ashdown took over again. By this time, the Conrero Alfa was dropping back to retire with transmission problems. Ashdown did his 30 laps in 2 hours 42 minutes and got the car up to 25th place before the Swiss resumed just after midnight. Gradually, number 45 started to slip down the field, and a piston blew after 18 hours when it was classified in 32nd place. Fastest lap went to Ashdown on the 16th tour at 5 mins 10.4 secs.

Graham Broadley had related in a letter to Ross before the race;

> "Mr Vögele's engine, which is an ordinary production unit and not specifically prepared by the makers, has already completed 11 hours at Sebring, 7½ hours at Nürburgring, many small races (actually two sprints and a hill climb) and is now headed for Le Mans, so keep your fingers crossed for us".

Obviously Ross didn't!

Overall, Paul Frère and Olivier Gendebien did the honours for Ferrari and Belgium.

* * *

Summertime... and the winning is (sometimes) easy

Chris Steele got his first win over Gordon Lee's Lister at Brands Hatch, while the others were away playing in Germany and Belgium, and did it again at the same circuit on 3rd July when he equalled the 57.8 secs class lap record. At Mallory Park on Whit Monday, Alan Rees lapped in 54.8 secs, which was fastest of the whole entry for the meeting, but was defeated by a loose petrol union in the race. On the same day, at a wet Crystal Palace, Ashdown and Steele battled until the latter spun at Ramp Bend but caught up again for a Lola 1-2.

The Hon Edward Greenall was getting the hang of his car and recorded several wins at BARC Aintree meetings on 18th June and again on 9th July when he broke the existing class lap record. Bill de Selincourt had the dubious distinction of winning what was said to be Goodwood's slowest race. The event was run in a torrential downpour, and Bill could manage only a fastest lap of a lengthy 2 mins 17.8 secs, but this was still enough to finish 47 seconds clear of the next 'swimmer'.

There was an International Sports Car race supporting the British Grand Prix at Silverstone on 16th July. Peter Ashdown was out in the works car, now rebuilt after its Belgian accident, and resplendent in a new red paint job with gold stripes. He,

Look out, behind you! The Hon Edward Greenall is monstered by Gerry Ashmore's Lister.

Ashdown, in the works car, leading Bruce Halford's Aston Martin DBR2 at the 1960 British GP meeting. Livery by now was a smart red with gold stripes.

Club Racing. David Bertram among the Lotuses and Coopers at Snetterton in August.

Rees and Greenall were in the up to 1500 cc class. In practice, Ashdown lapped in 1 min. 46.6 secs, which was good enough for the outside front row spot alongside Flockhart and Dickson's $2\frac{1}{2}$-litre Cooper-Monacos and Bruce Halford's 4.2-litre Aston Martin. The race was a cracker, with Ashdown making it up to second place behind Flockhart's Coombes-entered Cooper, although eventually dropping to third behind Douglas Graham's

2-litre Taylor and Crawley Lotus 15. Naturally, he won the 1500 cc class and lowered the class lap record with a time of 1 min. 46.0 secs—an average speed of 99.41 mph from only 1100 cc.

Over the August Bank Holiday weekend, Peter Ashdown again trailed the Coombes Cooper, this time driven by Roy Salvadori, at a wet Brands Hatch unlimited sports car race for the interestingly named Raffles Club Trophy: Rees was second to Ashdown in the 1100 cc class. Rees had faired better at the Nottingham Sports Car Club's Mallory Park meeting, lapping the entire field up to third place. David Bertram won the 1100 cc class at the BARC Aintree meeting, where John Wolfe was getting the feel of his new Lola, and finished behind the Scot.

Wolfe was fast but impetuous and, after finishing second to Lee's Lister at the Commander Yorke Trophy meeting held at Silverstone, lost two race wins at Mallory with spins. Wolfe continued with a win in one race and a spin in another at the AMOC Silverstone, and then had a tremendous scrap at Mallory with Rees and the Gilby. A win at the Peterborough MC Silverstone presaged the inevitable, when the Wolfe Lola arrived at Copse Corner far too fast on lap 6 of the BET Silverstone meeting and was extensively damaged. Alan Rees again won the 1500 cc class.

Typical August Bank Holiday weather! Peter Ashdown chases Roy Salvadori's 2½-litre Cooper Monaco at Brands Hatch in 1960.

The old prototype, 600 DKJ, had now found its way to Scotland, and Bill Mackay twice ran second to brother Jim's Lotus Eleven at Thornaby Airfield, near Middlesborough. As the season wound down, Bill de Selincourt won the 1100 cc race at the last Members' Goodwood, and then came second to Mike Salmon's Jaguar D in the unlimited sports car race.

Two new faces appeared late in the season. Lister-Chevrolet driver John Ewer sold his car, in what was then Southern Rhodesia, and bought a Lola Mark 1. He was desperately unlucky when he got pushed into the bank at his first outing at Oulton Park on 27th August, and, after having screwed it all back together again, he was eliminated in an accident during practice at the 10th September SUNBAC Silverstone, again through no fault of his own.

Chris Kerrison also hoped to get his hands on his new car by the BET Silverstone meeting, but it was the Boxing Day meeting at the Kent circuit before the car had an outing. Kerrison ran it with an FWB (1460 cc) motor to second place after stalling on the grid.

Ex Lister-Chevrolet driver John Ewer had little luck with his Mark 1 in 1960. He is pictured here at Silverstone at the North Staffs MC meeting.

Retrospective

All in all, a most satisfactory year for the Mark 1. By the end of September, Napier Road had built sixteen of the 1960 production cars (as well as nineteen Formula Juniors), and another batch was put in hand for Autumn build. The cars had proved very competitive in customers' hands, having recorded well over twenty class wins in the UK, international victories at Sebring and Nürburgring, as well as taking the US SCCA G Modified class by storm, as is illustrated in the next Chapter.

Peter Ashdown had again taken the BRSCC Sports Car Championship for the second year in succession, this time with Alan Rees a deserved runner-up in an older car. By the end of an eventful year, Lola Cars had built forty cars (production Mark 1s, Mark 2 FJs, an F2, and a Mark 3 rear-engined FJ): a tenfold increase on 1959!

The Lola bonnet badge as it first appeared on the works car at Oulton Park on 2nd April 1960.

CHAPTER 6

The Foreign Legion: Success in some unlikely Places

ALLAN ROSS's car was built alongside the other three 1959 cars at Maurice Gomm's West Byfleet premises. Ross actually had more experience with Climax-engined sports racers than Eric Broadley at that time. From early contacts he suggested a number of features that he wanted on his car for the US scene that he knew well, and some of these requests ultimately also found their way onto the later production cars. The roll-over bar which presaged the headrest has already been mentioned in Chapter 4, but one other innovation put forward by Ross as early as January 1959 was fibreglass bodywork, rather than aluminium. Broadley looked into this, but advice from the then young fibreglass industry was that anything less than a run of five or six bodies would be uneconomic, so Ross's 1959 car stayed metal.

Allan Ross: the Mid-Western pioneer

Ross wanted to take delivery of his car in late April 1959 but, things being pretty hectic at Parvis Road at this time, it was 26th May before the very first Lola landed in America at Chicago on the SS Pinemore. After a shakedown, the first race was at a small meeting at Wilmot Hills, Wisconsin. The Lola broke the class lap record and won by 26 seconds with a race time that was only fractionally slower than the winning Lister-Chevrolet in the larger sports car race. A much more serious test was to come at the Road America June Sprints at Elkhart Lake. Ross again saw the chequered flag first at the Road America Circuit and, in the course of the race, became the first G Mod car to break the 3 minute barrier at the 4 mile track. Two further wins were added at Buckley Field, Denver, Colorado on 12th July, and two more at Stout Field, Indiana. This latter event had a guest overseas driver in one race—Eric Broadley! Eric's sister lived in Toronto and he combined a quick trip to Canada with a meeting with Ross to sort out the latter's position as Lola's first agent in the US. Ross had christened his light blue car 'Lulu', and swept all before him that first season. Perhaps the most important win came at Watkins Glen on 26th September. No less than 33 under 1100 cc cars started at the Finger Lakes circuit, and the single Lola won from Ed Crawford, in a Briggs Cunningham OSCA, and the Elva Mk V of Chuck Dietrich. 'Lulu' knocked 3 seconds off the Watkins Glen lap record, and recorded her tenth win in ten starts.

Naturally, by this time, Ross was besieged by fellow-countrymen wanting to know how they could get their hands on the new English sports car.

KENILWORTH, ILL.
727 CUMMINGS AVE.
ALPINE 1-3455

HOMEWOOD, ILL.
SYCAMORE 8-9070

LOLA-AMERICA

UNITED STATES AGENT

A. M. ROSS

SPORTS RACING
FORMULA JUNIOR

"Lola-America"

Following Eric's July visit, Ross had set up as 'Lola-America', and started to take enquiries from all over the States for more cars, particularly after a number of articles about the new car appeared in the American specialist press at the end of 1959. The first two cars he imported from England were sold in California to Art O'Connor, and Ron Hathaway.

Art O'Connor was a Flying Officer in the US Air Force, stationed at Edwards Air Force Base. He was well known in Californian racing circles, and had previously bought one of the 1959 ex-Team Lotus Elevens from US importer Jay Chamberlain. However, when Ross appeared with his Lola, O'Connor disposed of the Lotus and ordered a new car from the Chicagoan.

O'Connor's car was originally intended to be aluminium but, by the time it was shipped in February 1960, fibreglass bodies were all the wear. Such were the problems with delivery of engines from Coventry Climax around this time that the car was shipped without this component, which had to be air-freighted out later. First time out at Riverside was not very auspicious as the bonnet flew off at some speed when attacking an RSK Porsche into Turn 9. The factory got blamed for this drama but, in fact, it was self inflicted. Air temperatures in California were a bit higher than any Mark 1 had experienced up to that point, and O'Connor had some overheating problems in practice. The solution had been to cut away the aluminium duct behind the radiator, thus exhausting all the hot air straight under the bonnet, rather than under the car, to the detriment of the bonnet springs.

When repaired, O'Connor ran the car at Pomona and picked up a win at Laguna Seca, before selling the car on to LA Mobile Home manufacturer, Art Snyder, for 1961. Snyder raced the car quite often at West Coast circuits in that year, recording G Mod wins at Las Vegas and Laguna Seca.

Art O'Connor's second outing, at Del Mar, CA in May 1960, showing the minor damage from the previous Riverside meeting when the bonnet flew off.

(above)
Ron Hathaway chases John Miller's Lotus Eleven.

Hathaway, meanwhile, had to wait until June to get his hands on his car, and did not really feature much in 1960 apart from a third overall and class win at Laguna Seca at the end of the season. The following year he took a win at Cotati, but had to give best to Snyder in G Mod at the mid-season Monterey meeting.

Allan Ross was becoming deeply involved with his new restaurant in early 1960, which had become a meeting place for enthusiasts in the Chicago area. Not only was this cutting into his possible racing time, but he was also under pressure from potential Mark 1 purchasers to a greater extent than the factory

could expect to deliver. One who exerted the most pressure was a driver from New York State called Millard Ripley. Ripley had some experience, including a run at Sebring in 1959 in a Lotus Eleven, entered by Jay Chamberlain. By April 1960, many telephone calls and increasing amounts of greenbacks, persuaded Ross to sell 'Lulu' to the Ithaca motor trader. Millard Ripley was a pretty hefty individual, and he had uprated springs fitted to the car before racing it. Extra weight or not, he set his sights on the SCCA G Mod class title for 1960, and duly achieved this objective. At the long distance race at Elkhart Lake, Ripley shared the car with its previous owner and future USAC man, Burdette Martin, but they ran second to an ex-David Hitches Mark 1 just imported direct from England by Jessie ('Doc') Wyllie, and driven by him, his wife, Peggy, and... Ripley. The imported car was picked up from Kennedy airport, given a fresh coat of paint in Pittsburg and, according to Wyllie; "won Class G handily on English water and oil which had not been drained on shipment!" Presumably Ripley scored points for both finishing places.

A follow-up success at the 1961 Sebring 12 hours

Two further cars were sold by Ross in 1960, and delivered in early 1961. They went to Chuck Hall, brother of Jim Hall of later Chaparral fame, and Charlie Kurtz, who ran a sports car business in Clearfield, Pennsylvania. The Kurtz car had been on order for a long time, but the Hall car had been finished for display at the 1961 BRSCC Racing Car Show, and then had a brief spell in Ashdown's showroom, before being shipped to America. Both cars were prepared for the Sebring race with twin tanks, a 4.55 axle, a twin plate clutch and all the appropriate Appendix C equipment, such as a tall windscreen and luggage compartment modifications. A last minute panic occurred when the cars were dispatched without hoods. Unfortunately, the upholstery firm making them had a fire and lost both the hoods and the pattern, and a single emergency version had to be made quickly in the States to swap surreptitiously between the two entries to keep the scrutineers happy!

The two cars had been entered by Lola-America for Chuck Hall and Ross, and Charlie Kurtz co-driving with Millard Ripley. Both the Lolas ran steadily, with the Hall/Ross car

(opposite page) West Coast racing in the '60s. Art Snyder is sandwiched on the front row by the Lotuses of film director Kurt Neumann (32) and the ex-works 1958 Sebring car of Frank Monise (44). Just visible (73) is Pete Brock, future designer of the Shelby Daytona Coupés.

Wonder how it handled on corners?

These two pictures show Millard Ripley's remarkable 'transporter' being loaded in the paddock at Bridgehampton, NY in 1960, and in action below.

leading the 1100 cc sports car class after 6 hours—but then dropping back with a small problem, handing the class win to Kurtz and Ripley. Phil Hill and Olivier Gendebien wrapped up the overall win for Ferrari.

Kurtz's Sebring class winning car ran throughout 1961 mostly at East Coast events, doing battle with 'Doc' Wyllie and Bill Wilbourne, who had bought the ex-Ross, ex-Ripley, car. In the final SCCA points reckoning, Kurtz just shaded Wyllie, with

The Chuck Hall car, photographed at Napier Road, after preparation for the 1961 Racing Car Show.

Wilbourne fourth. At the end of the year Kurtz sold the car to Art Tweedale. Sadly, Kurtz himself was killed in a road accident in 1964.

An OSCA engine for Oscar

Oscar Schmidt was another of the Mid-West sports car racers known to Ross. He ran Import Motors of Chicago, who were the area Volkswagen distributors. He had run Italian OSCA sports cars in local events and, seeing Ross's early success with 'Lulu', had approached him with the idea of putting an OSCA engine and gearbox into a Lola chassis to run in the smallest SCCA H Mod class.

By 1947 the three Maserati brothers had fulfilled their contract with Commendatore Orsi and set up their own 'Officine Specializzate per la Construzione di Automobile – Fratelli Maserati SpA; which, fortunately for the size of the car badge, shortened to OSCA. They built some excellent sports cars in the 1950s, initially based on a single overhead cam head on a Fiat block, but later with their own twin-cam, four-cylinder engines, which varied between 747 and 1491 cc. The cars sold quite well in the United States, particularly after Bill Lloyd and Stirling Moss had played tortoise to the Lancia, Ferrari, Aston Martin and Cunningham hares, and won the 1954 Sebring outright with an under 1500 cc Briggs Cunningham owned car. OSCA had also made the middle and small classes of the Mille Miglia their own in the mid-1950s, but Porsche were taking over the 1500 cc class, and the 1100s were starting to come under threat from the Climax-engined cars. The brothers were now well over 60, and decided to put their effort into the 750cc engine for the sports racing class.

Schmidt himself visited Bromley in October 1959, and the plan was for him to arrange for a 750cc OSCA engine and gearbox to be delivered to England for fitting into a Lola chassis. There were no drawings for the engine and, although work progressed on the car, the factory were dependent on the delivery of the power unit. Because of a delay in the British Customs, Lola did not get their hands on the engine and gearbox until the end of January 1960. Although the engine was a jewel, its installation was clearly very different from the normal Climax. In the event, fitting it in proved so difficult that the original chassis built for Schmidt had a Climax engine installed and became the car Ross sold to Hathaway. Bromley started again with a completely new chassis that this time could be built up specifically for the OSCA engine. When the car was eventually completed, Eric took it out for a run on the A21 for a test without bodywork, and with the connivance of the Kent County Constabulary! All of this much delayed things, and Schmidt had to soldier on with the old OSCA car until the Lola-OSCA ran at Elkhart Lake at the season's end. The wait was well worth it, however, and Schmidt proved almost unbeatable in H Mod in 1961, and won the SCCA Championship.

Schmidt sold the car to John Holmes, for himself, and dye company owner, Guy Bates, to drive. In 1962 Holmes finished fourth in the National points standing, and Bates placed sixth in the North East. By now the H Mod class was at 850ccs, and the Lola was running short of some Italian horses, so plans were put in place to add another 100cc. Bates still only scored places rather than wins, and ended the 1963 season in fourth place among the H Modifieds.

East Coast G Men

Mention has already been made of Gulf Oil executive 'Doc' Wyllie and the ex-David Hitches car. Following the win in the 500 mile race at Elkhart Lake in September 1960, Wyllie continued with class wins at two regional meetings the following month. In 1961 Wyllie achieved nine class victories in the car, although just missed out on the SCCA national championship when he had to pass up a critical Bridgehampton race to drive a Lotus Elite at Le Mans. Wyllie was so impressed with the car that he ordered one of the Girling disc braked 13-inch wheel 1962 cars from Bromley. For a time he ran both cars, with his wife Peggy driving the older model. The newer car remained unbeaten in 1962 until a rod broke at the September Watkins Glen meeting, but the earlier results were sufficient to secure Wyllie the G Mod title for that year.

For much of 1962 and early 1963 Wyllie's main opposition came from Art Tweedale with the ex-Kurtz car. Wyllie won with his older car at Marlboro, but Tweedale replied with a class win at Virginia International Raceway. It continued like this for much of the season, with the oil man winning at Cumberland, Stuttgart, Connellsville, Thompson and Meadowdale; and Tweedale following up his early success with victory at the Road America long-distance race at Elkhart Lake, with Ed Baptista sharing the driving. The same two drivers continued their duel at the beginning of 1963, with Wyllie placing second in the final

Sebring 1961. Kurtz/Ripley class-winning car, with the owner behind (in the jacket). The nose of the Hall/Ross car can also just be seen.

Charlie Kurtz at Thompson, Connecticut in 1961.

SCCA national points list for that year. For 1964 Wyllie returned the car to Bromley for fitment of a Lotus Cosworth 1600 cc twin-cam engine and Ford gearbox, with which to contest the 2-litre class, although he still retained the smaller Coventry Climax engine for some races.

Down in the South East, Roger Heffler and the Miami Castle Racing Team had one of the later cars which clocked up ten wins in twelve races, including one at Nassau.

Big accident at the Canadian GP

One of the 1960 cars had gone to the Hollingshead Company in Toronto, Canada to be raced to advertise their range of 'Whiz' car care products. It was bought later that year by Ross de St. Croix of Montreal, who ran it with some success before entering the Canadian GP at Mosport in 1962. There he had an enormous accident as a result of contact with Jim Hall's Chaparral during a rain storm, and was lucky to survive.

Road and Track reported at the time:

> "The ambulance careened onto the course a couple of times. First sortie was for Ross de St. Croix, whose Lola flip-flopped off the track at the beginning of the back straight. He somehow managed to miss a stand of trees, a pile of

rocks, and a culvert, and got away with his life, if not all of his skin. The car was a shocking mess, looking as if a large someone had tried to bend all four comers up at the same time."

David McKay down under

Australian David McKay came to Europe in 1955, and he and his partners bought three customer Aston Martin DB3Ss and ran them under the 'Kangaroo Stable' banner. The most notable result he achieved was a second place in the Hyeres 12 hours with F O A (Tony) Gaze assisting with the co-driving. McKay's car was shipped back to Australia, and replaced in 1957 by another DB3S, this time an ex-works team car that Moss/Collins had taken to second place at the 1956 Le Mans. McKay had the car painted red, and took a number of wins in Australia with the Aston over the next two seasons.

An Australian journalist friend in the UK, who had navigated with McKay on the Liege-Rome-Liege and Tour de France, saw one of the Lola's sensational outings at Brands Hatch, and wrote to David urging him to buy a car. In fact, it was the 1960 works car that McKay saw on a visit to Bromley and wanted in mid-

David McKay's little spending spree. The new owner proudly sits in the ex-works Mark 1 with Bob Atkin in striped jumper behind. Tom Corcoran, in the Mark 2 Formula Junior, was later to buy it to prevent McKay's bank balance from blushing.

season. He persuaded Graham and Eric to agree to give him first refusal, in spite of pressure from from Allan Ross—who offered more money to get the car to the US. Perhaps the reason was that, as he said later, his enthusiasm for the marque somehow got ahead of his bank balance, and he offered to take a 1960 Formula Junior too!

But a financial crisis was on the horizon, in spite of the sale of the DB3S, as David McKay recalled recently in the Australian magazine *Motor Racing Australia*:

> "Before their (the cars) arrival I had heard the Government had jacked up the sales tax on cars to 40% and thought nothing of it until I realised it applied to racing cars too. The forwarding agent, TKM, informed me it would like a cheque for Aus£7,775 before releasing the cars, and that galvanised me into deciding the Junior had to be sold at once, not even run for fun first.
>
> Fortunately, Sydney enthusiast, Tom Corcoran, wanted the Junior and had Aus£3,000. He came to my Wahroonga home, fell in love with the mini Vanwall-shaped toy, and the financial crisis was averted."

McKay was much impressed with his new mount, and noted later how the quality of workmanship on the car, and the support he got from the factory, was in an altogether different league from what he had come to expect from Feltham with his Aston Martins. His first run with the car was in November 1960 at the newly constructed Warwick Farm track, which had been built round a horse racecourse in the suburbs near Sydney. The Lola was one of the very first cars to lap the new circuit in testing. The circuit opened proper on 18th December 1960, when McKay recorded fastest lap in finishing second overall and winning his class. The Lola continued with class wins at Ballarat, Albury, Longford and Bathurst, although it usually had to give best for outright victory to Matich's Lotus 19 and Stillwell's Cooper Monaco, both of which had 2.7-litre Climax FPF engines.

It was in an effort to keep up with the big engined Climax cars, and compete in the 1961 Warwick Farm GT championship, that McKay took the car to Clive Adam's Sydney panel shop for a fastback hardtop to be added. It was a disaster, visually and

David McKay's hardtop was to prove as ineffective as it was unlovely.

technically. The 50 lbs of aluminium, added high to the rear of the car, did not do wonders for the handling, and lap times were well down on what had previously been recorded with normal bodywork. Nevertheless, McKay did finish second to Matich's converted Jaguar D-type in the GT race at Warwick Farm on 30th July 1961, and won the class.

Third in the Warwick Farm race was a Porsche driven by Canberra rally driver, Greg Cusack. Cusack was to buy the Lola from McKay later in 1961, and subsequently removed the ugly hardtop in a roll at Catalina Park when a rear shock absorber shaft broke. While in Cusack's ownership, Chris Amon drove the car to a win in the Victoria Sports Car Championship at Sandown on 16th September 1962.

* * *

Motokar Sport in Malaysia

One of the last cars built at Bromley in 1960 was fitted with a 1220 cc FWE engine, and sold to Chan Lye Choon of Singapore. Chan Lye Choon was one of the most successful of Malaysian drivers who competed on what we would now call Pacific Rim circuits. He and his elder brother, Chan Lye Huat, who owned Citroën agents Eastern Autos, both started racing in Singapore before the Japanese occupation. They resumed in the early 1950s with a pair of MG TDs, and moved on to a V-twin powered Cooper, before Chan Lye Choon bought a production Aston Martin DB3S. With this he scored a number of sprint and hill climb successes, and a win in the 1958 Macau GP.

Fine signed and dated photograph of Chan Lye Choon winning the 1961 Jahore GP.

Just like McKay in Australia, he was immediately much more at home in the better handling Lola than the unwieldy Aston. His most notable success was in the 1961 Jahore GP, although it has to be said his opposition was far from outstanding. Once Cowling's Cooper F2 ran into trouble, the only possible threat was from Peter Brown's Lotus 17, and Chan Lye Choon was able to cruise to an easy win by several laps in the 145 mile event to collect his trophy from the Sultan.

By 1963 Chan Lye Choon had replaced the Lola with a Lotus 23B for sports car events. The Lola passed through several pairs of hands, but still kept on competing in local events until the early 1970s: something of a testament to the car's staying power in an area where skilled race maintenance cannot have been plentiful.

... and New Zealand

Two of the 1961 cars made the long trip to New Zealand. One was fitted with an FWE engine, and supplied to J S Cottle. He ran in the Renwick '50', finishing third in both the Formula Libre race and the sports car event, after leading the latter when the throttle linkage broke. In 1963 Cottle qualified the car for the New Zealand Grand Prix at Pukekohe with a lap of 1 min. 40 secs, which compared reasonably with Bruce McLaren's Cooper at 1 min. 26.8 secs. In 1963 he also became the New Zealand Sports Car Champion.

The other car went to Dave Lawrence of Wanganui, without an engine. Lawrence fitted his own Climax 1500 cc engine from his Cooper Bobtail, but this engine blew up terminally shortly afterwards at a race at Ohakea, and was replaced by a similar-sized Ford push-rod unit. Lawrence used the car to place third to Cottle in the 1963 Championship.

The two New Zealand cars continued to do battle in the hands of new owners for several seasons in the Sports Car Championship. In 1964 rival motor traders Riley and Dawson replayed the earlier encounters, and a year later this continued with Bill Caldwell and Donald McDonald. Eventually the Ford-engined car gained a Lotus twin-cam head, Brabham 7-inch wheels and disc brakes when it was in Dave Oxton's hands in 1966.

* * *

Mark 1s in New Zealand. Bill Caldwell (31) just leads the Ford-engined car of Donald McDonald at the NZ GP at Pukekohe in January 1965. It was to all end in tears shortly when McDonald, having got ahead, lost it and was collected by the closely following second Lola.

... and South Africa

Lola Mark 1s first made an appearance in South Africa in the 1962 Rand *Daily Mail*-sponsored 9-Hour Endurance race at Kyalami when Englishman Jack Paterson and Bill Ferguson from Johannesburg shipped their cars out from the UK for the event. Early on, the Lolas lay fifth and sixth but both dropped back with various problems. At the end, the Paterson/van den Broeck car had got back up to fourth place, but spun on the last lap and ended up seventh.

The South African owned car passed to Henri Le Roux, and he entered it in the 1963 9-Hour race with Gert Coetzer co-driving. The forty-one entrants, as always, were a mixture of saloons, GTs, production sports, and some sports racers. The race ran with both overall placing and handicap against a set distance, and hence average lap time. There were also specific classes too, although the Lola only had a couple of Austin-Healey Sprites and a DB Panhard Le Mans to contend with.

Quickest in practice was a local Lotus 23 Alfa, but this crashed. The race itself was a duel between the 1962 winner, David Piper's Ferrari GTO, co-driven with Tony Maggs, and the Willment AC Cobra of Bob Olthoff and Frank Gardner. The Lola initially ran fifth behind the two front runners and David Prophet's Lotus 23B, and a locally driven Porsche Spyder. This became a fine third place at the finish when the German car crashed and burned, and Prophet went out with a dropped valve. The Lola fared less well on handicap, and came in seventeenth and last. The organisers seemed to have been reading about the car in the British press, and set a handicap lap time 2 seconds a lap quicker than the Cobra, and 5 seconds faster than the GTO!

The car again showed up in the 9-Hours in 1964 with Coetzer and Podmore driving, but could not repeat the previous year's success.

Not bad for a new car manufacturer

The wide geographic spread of success of the Mark 1, that had only emerged from the Bromley lockup a couple years before, was quite remarkable. Undoubtedly, such early broad exposure of the new marque's name contributed to its growth in subsequent years.

CHAPTER 7

The Mainstay of the Company in 1961

BY 1961 pure sports car racing was starting a slow decline in Europe. At club level, the popularity of GT and production sports car racing was proving a counter attraction. Internationally, the move was towards GT racing, evidenced by that year's Le Mans, which had 32 closed cars out of the 55 starters. On the driver front, Formula Junior had really taken off, and the up and coming men were all to be found in this competitive single seat formula. Nevertheless, there was still plenty of opportunity to race a sports racer, and the Mark 1 continued to provide the class of that competition.

For Lola Cars Ltd this was just as well. The Mark 2 Formula Junior introduced at the very end of 1959 had not been the outstanding success that might have been hoped for, given the sports car pedigree. Perhaps it was too easy to rush out the car on the back of the Mark 1's impressive achievements. Certainly the power unit was at the wrong end for this most competitive of environments, although Lola suffered also from poor engines in the beginning of 1960, without the early access to Cosworth Engineering that the rival Lotus 18 had. Dick Prior and Peter Ashdown ran the works Mark 2s and could generally give all but the Team Lotus 18s a run for their money. By August, when some of the engine power and reliability problems had been resolved, Prior picked up a win at Snetterton in the *Motor Racing*

National FJ Championship, in which he and Ashdown were to finish sixth and seventh to joint team Lotus winners Trevor Taylor and Jim Clark. Other Mark 2s were run by the quasi-works Fitzwilliam team with a variety of drivers. John Love won for them at the Belgian circuit of Chimay but, generally, they were second or third to the Lotuses or Coopers. All but two cars had Ford 105E engines, although local Eltham man, Dennis Taylor, performed wonders with a BMC version. In all, nineteen of the model were sold, plus a one-off 'mirror-image' F2 car with a home-brewed Gerald Smith twin-cam head on an FWB Climax.

For 1961 the sports car remained in production in unchanged form and it was joined by a new Formula Junior, the (now officially designated) Mark 3. This was Eric's first rear-engined car, and fairly bristled with original ideas. The driver sat well forward with the battery between his knees and the fuel tank behind the seat. The body was a strange but attractive boat shape, mounted quite high. The Mark 3 was not a success, and handled a bit like a ship too! Only eleven were produced and as Eric himself recalled on the car in an interview published in *Motor Sport* in 1989:

> "It was not a particularly good move. Write that one off to experience."

Hence, without the stability of continued production of the Mark 1, Lola Cars might not have survived another year. The second win of the new season, however, was scored by a very unusual Mark 1.

* * *

The UDT-Laystall Lola

During 1960, Eric Broadley had been approached by Basil de Mattos of Laystall Engineering, Great Suffolk Street, London SE1 to build a car for finance house United Dominion Trust (UDT) with an FPF Coventry Climax engine developed for 1961 F1, and to be supplied by them. The car was planned around a more robust version of the Mark 1 chassis. It had Girling disc brakes all round, adjustable shock absorbers, dry sump lubrication for the FPF, and a large combined oil cooler/radiator. Suspension

was virtually standard Mark 1, but made from heavier gauge tube. The rear driveshafts and stub axles were similar in design to the Mark 1, but featured larger shafts and universal joints. The front hubs were specially turned from solid steel. Wider section wheels and tyres were fitted. The gearbox was a rear mounted

The FPF-engined UDT-Laystall car shows off its bulbous rump while sitting in the Goodwood paddock stalls. It was later to win its only race on UK soil in Henry Taylor's hands.

five-speed, designed and constructed by Laystall themselves, and it was a pretty massive affair built with F1 in mind. A normal Borg and Beck twin plate clutch was fitted.

The body was in aluminium, with a high tail to clear the large tanks required to supply fluids. There was a bigger blister in the bonnet for the engine, and a lump on the instrument binnacle for a rev counter.

By 1961 the UDT-Laystall team had, to some extent, lost interest in the project. They had bought three Lotus 19 'Monte Carlos', usually fitted with 2½-litre Coventry Climax engines, and driven mostly by Cliff Allison, Henry Taylor and Graham Hill — although Stirling Moss, Mike Parkes and Innes Ireland were also seen on occasions. Eric tested the Lola, but the one and only British outing was at the 44th Goodwood Members'

Meeting on Saturday 11th March 1961 when the car was driven by Henry Taylor.

The idiosyncratic Bill Boddy reported in *Motor Sport*:

> "A 10-lap Scratch race for Unlimited Sports Cars was extremely good value, because three big cars with 3.8-litre Jaguar engines ran, and it takes the big fellows to really stir the onlookers — the present tendency in all branches of motor racing to reduce engine size kills the spectacle only powerful engines can supply. Yet here we saw a 1½-litre car beat the heavy metal when, with consummate skill, Henry Taylor, trying out the new Laystall five-speed gearbox in twin-cam Lola virtually akin to an F1 car, crept up on de Selincourt's Lister-Jaguar and, the latter missing a gear change during this side-by-side tussle, won by 2.4 sec. It was a truly exciting race..."

The de Selicourt Lister was the original Peter Whitehead car that had been driven the previous season by John Bekaert, who himself was just about to embark on a successful period with a Lola.

After Goodwood, the Laystall car lay unused for the rest of the year until, now shorn of its original green paint job, it was sold through Bromley by Alan Ross to Bill Stone of Chicago. In the States it appeared fitfully in F Modified, entered as the 'Lola Medallion', where its best finish appears to have been a fourth at Meadowdale in September 1962.

Stone advertised the car for sale in September 1964:

> "... Climax DOHC engine, needs one rod and piston!"

The car was later still fitted with the 1100 cc FWA engine from Ross's 1959 car.

* * *

The class of '61

Over fifteen drivers were seen in Mark 1s in 1961. Among the more successful were Peter Boshier-Jones (who had bought the Hon Edward Greenall's car); Laurie Keens; John Wolfe (when he was able to stay on the black stuff); and Ian Harrison-Hansley. Chris Kerrison ran some good races, as did John Ewer and Nairn Ferrier in Scotland. One particularly welcome visitor to Bromley to buy his new car towards the end of 1960 had been Lord Angus Clydesdale, who would always take everyone for beer all round in the Stanley Road pub! Someone swapping brute force for elegance was John Bekaert. Bekaert's car was owned by Northamptonshire businessman, Derek Wilkinson, who had also bought the Lister they had run the previous season.

Although the UDT-Laystall Lola had had a win at the first race meeting of the season, the first actual race at Goodwood on the 11th March was won by Peter Boshier-Jones from Laurie Keens—who recorded fastest lap. As was usual, a huge crowd turned up on Easter Monday at the same venue, where John Bekaert finished sixth in the 15 lap unlimited sports car race for the Sussex Trophy behind the 2½-litre Climaxes and John Coundley's Lister-Jaguar.

Hill Climber. Peter Boshier-Jones was an accomplished circuit driver, but also took to the hills on occasion. This excellent shot was taken at Rest and Be Thankful.

John Bekaert's Derek Wilkinson-owned car in the Silverstone paddock.

The Snetterton MRC opened their first meeting with two unlimited sports car races, both of which were won by Bekaert from Ian Ballie's Aston Martin DBR1. John Bekaert held off Chris Kerrison and Irishman Tommy Hayden, in the old Bernard Cox 1959 car, to win again at the BARC Oulton Park meeting on 15th April. Stirling Moss won the Unlimited sports car race at the BARC Aintree 200 meeting with a UDT-Laystall Lotus 19, while in the 1150 cc class John Wolfe did well to bring his clutchless

Bekaert again, this time holding off Chris Kerrison's car at the 1961 International Trophy meeting.

Mark 1 home first, ahead of Kerrison and David Hitches (out in his new 1961 Lola that replaced the earlier car sold to 'Doc' Wyllie in the US).

Ian Harrison-Hansley picked up a couple of wins with his new Mark 1 at the Jaguar DC event at Brands Hatch, and the Maidstone and Mid-Kent MC at Silverstone. Back at Goodwood for the next Members' Meeting on 6th May, Boshier-Jones and Keens repeated the March result when the latter was penalised for a jumped start.

With the British manufacturers still grumbling about the new 1½-litre Formula 1, the 1961 International Trophy at Silverstone was run for so-called Inter-Continental Cars. Stirling Moss was absolutely superb on a wet and oily track in Rob Walker's Cooper-Climax T53 2½-litre, lapping the entire field—which included Messrs McLaren, Brabham, Hill, Ireland, Surtees, Salvadori, Brooks, Clark, Allison, and a few others who could peddle a bit. In the opening 25 lap Sports Car race, the smaller class was up to 1500 cc and this was won by Henry Taylor's FPF-engined UDT-Laystall Lotus 19 from the dicing Lolas of Bekaert and Kerrison. John Wolfe was a pits' visitor with an oil leak.

Silverstone on the 13th May saw the Martini International Club Trophy being held for sports cars. The up to 1300 cc class was won by John Ewer, while Brenda Dickinson suffered a loose bonnet on the ex-Alan Rees 1959 car. Peter Boshier-Jones showed his hill climb background by winning classes at Loton

Mrs Brenda Dickinson in the ex-Peter Gammon/Alan Rees 1959 car.

Park and the Bugatti OC sprint at Silverstone. Laurie Keens added another win at the Snetterton MRC Stanley Sears Trophy meeting, with Angus Clydesdale close behind.

During 1961 MGM were making a film of a novel called *The Green Helmet* starring Syd James and Bill Travers. The Whit Monday BRSCC meeting featured an unlimited 15 lap sports car race, labelled the Green Helmet Trophy, and with pots and prize money put up by the film company. The UDT-Laystall team of Lotus 19s were using the event as a tryout for the following weekend's Nürburgring 1000 Kms and had especially fitted some knock-off wheels. Red faces all round when Henry Taylor and Mike Parkes both lost wheels, and even more so when Roy Salvadori's Cooper Monaco picked up the win for rival finance house Yeoman Credit Racing. In the smaller 1600 cc class, John Wolfe, now with a larger FWE engine, collected MGM's money ahead of Ian Harrison-Hansley.

* * *

Brake lights on in the rain for the Kerrison/Sargent class-winning car at the 1961 Nürburgring 1000 Kms.

1-2-3 in Germany again

Sixty-three starters faced the flag for the traditional 44 laps of the Nürburgring 1000 Kms. They were in seven classes, four of which were for sports cars and three for GTs. Class 4 (851-1150 cc) had four Lolas for last year's winners Vögele/Ashdown, Bekaert co-driving with de Selincourt, Chris Kerrison and Peter Sargent, and Hitches with his new car driven again with Bob Hicks. They were opposed by two Austin-Healey Sprites and two works DB Coupés.

Before the race the FIA had attempted literally to clarify the position on the tall Appendix C windscreens for World Sports Car Championship events, since those using Perspex had been having visibility problems. As so often happens, the amendment caused even more confusion and allowed for a slot 35 cms wide to be cut in front of the driver: unfortunately, it did not say how deep the slot should be. The result was that most of the sports cars with Perspex windscreens, including some of the Lolas, hacked out a complete chunk.

In the dry Thursday practice the Kerrison/Sargent car was the quickest of the Lolas, some 19 secs ahead of the Swiss car. In the race, Ashdown had the Charles Vögele car into the class lead initially, ahead of Hitches/Hicks, and Kerrison/Sargent (who were struggling with a broken rev counter). After the pit-stops on lap 16, first rain and then snow began to fall, making things

June 1961. Rouen GP. David Hitches' class-winning car before the race.

pretty nasty for everyone. By half-way, the Kerrison/Sargent car was in front of the 'Lola class', from Hitches/Hicks (which was soon to retire), and the Swiss car. At the finish, Chris Kerrison /Peter Sargent finished in 14th place overall to win the class from Bekaert/de Selincourt in 17th, and Vögele/Ashdown a place behind.

The Rouen GP, where the race was run in two heats, each of two hours, provided David Hitches with some consolation for his German disappointment, when he won his class in his last outing with the car.

From Dunboyne to Charterhall

The second half of the British season carried on much as the earlier part. Boshier-Jones had a purple patch with victories at NSCC Mallory Park, an unlimited sports car race at the BARC Members' Aintree on 17th June, and on the hills at Prescott and fastest sports car at Shelsley Walsh. Peter Dickinson won at the 46th Goodwood BARC members meeting on 10th June, and Keens did the same thing at the July meeting. John Wolfe started from the back of the grid at the Midland Motoring Enthusiasts' Club Silverstone meeting, but still won the under 1500 cc race from John Ewer's FWE engined Mark 1. Ian Harrison-Hansley took the class at the BRSCC Silverstone meeting on 29th July, although both Mrs Dickinson and Laurie Keens retired. August Bank Holiday Monday saw most of the cars at Aintree for the BARC meeting where John Wolfe won both his heat and the final, in spite of a spin in the latter.

Curiously, the first ever major race meeting held in the British Isles took place in Ireland in 1903 at Athy, near Dublin, as road racing was banned on the mainland. Ireland continued this tradition of holding races on the public highway, and the main event of the Leinster Motor Club meeting on 29th July 1961 at Dunboyne was the 25 lap Leinster Trophy Formula Libre Handicap. With three laps to go, Tommy Hayden's Lola held a 40 sec lead over Gerry Ashmore's fast closing F1 Lotus 18. Two laps and the gap was down to 26 secs, and then 16 secs as they entered the last tour. Dunboyne village erupted in cheers as the white Irish Lola held on to beat the Englishman by 3.8 seconds: local Guinness sales were later to hit record proportions!

Ian Harrison-Hansley leads John Ewer at a Maidstone and Mid-Kent race meeting.

The West Essex Car Club meeting at Snetterton on 20th August was a good example of the situation existing in small sports car racing towards the end of 1961. The opening 10 lap race drew twenty five entries, of which seven were Mark 1s. Wolfe non-started, but Peter Boshier-Jones showed class to score a fine close win from Greene's Gilby. John Bekaert was third, after a slight coming together with the other two, only just in front of a train consisting of Clydesdale, Keens, new driver John Nicholson in the ex-Hitches car and entered by Farnborough Racing Enterprise, and Harrison-Hansley.

Nice weather greeted the competitors at Crystal Palace on 1st September, where Boshier-Jones scored a runaway win from a scrapping Harrison-Hansley and Nicholson, and rounded off his successful season with a further victory at the Brands Hatch GP circuit during the concluding BRSCC meeting. Brenda Dickinson scored wins at both the Peterborough MC and North Staffs MC Silverstone events in September. Lolas were 1-2-3 at the last BARC Members' Meeting at Goodwood on 16th September, and Nairn Ferrier beat Lord Clydesdale in the 1300 cc class at the final Scottish meeting of the season at Charterhall.

The Mark 1A: an attempt at a Clubman's Mark 1

Just before Christmas 1961, Lola Cars announced a down market version of the Mark 1, perhaps to spin out production a little longer and use up stock components. The car was to be sold in kit form in the UK or assembled for export, due to the high cost of shipping a crated knock-down version.

The basic specification came with a standard tune Ford 105E or 109E engine, pressed steel wheels, Firestone Sports Nylon tyres and 9-inch cast iron drum brakes. Base price was £965. From there, a huge range of optional extras and additions were offered, including Climax- and Ford-tuned engines (just as well with the standard Anglia giving a manly 40 bhp on its single Solex!), various gearboxes, disc brakes and so on.

In fact, no actual production cars were sold but the prototype Mark 1A, fitted with disc brakes and a decent Ted Martin 109E three valves per cylinder engine, found its way to Canada in 1962. In 1963 it ran in the 3rd Canadian GP, which in those days was for sports cars, at the Mosport circuit near the shore of Lake Ontario.

Overview of 1961

In all honesty, there was not much opposition to the privately run Lola Mark 1s in 1961, with only Keith Greene's Gilby and talented Mike Beckwith's old Lotus Eleven occasionally disturbing the Bromley bandwagon. However, things were about to change. Lotus, their roots founded in sports cars, were in the business of replacing the disastrous 17. Other old rivals, Elva, went bankrupt early in the year, but re-emerged as Elva Cars (1961) Ltd at the original factory, and were also about to produce something new for the small sports racing car market.

Lola Cars had made about six Mark 1s, and the Mark 1A during the year. They had been busy sorting out the recalcitrant Mark 3 Formula Junior, and designing the vastly improved Mark 5 for the same category in 1962. Eric was also designing the Mark 4: Lola Cars first venture into Formula 1. John Surtees was to be the lead driver for this effort, with Roy Salvadori in the second car, and Reg Parnell as Team Manager. Funding came from Yeoman Credit, although the team actually became known as Bowmaker Racing at the beginning of 1962.

LOLA SPORTS RACING CAR 1962 - 3

Price List - Ex Works, Bromley, Kent, England.

Standard car in component form with Ford Anglia 105E or Classic 109E engine & gearbox in standard tune, pressed steel wheels, Firestone Sports Nylon tyres and 9" drum (cast iron) brakes.	£965. 0. 0.
Additional for higher grade specification as follows:-	
Al-fin aluminium brakes, 10" front, 9" rear, mag. backplates.	+ £ 30. 5. 0.
Girling disc brakes, 9½" front & rear, light alloy calipers.	+ £ 57.15. 0.
Magnesium road wheels in place of pressed steel (4)	+ £ 69.14. 0.
Dunlop R5 Racing tyre & tubes in place of Firestone Sports.	+ £ 16. 0. 0.
4-Speed close ratio gears for Ford gearbox. (Exchange-unfitted).	+ £ 35. 0. 0.
5-Speed close ratio gears for Ford gearbox. (Exchange-fitted).	+ £120. 0. 0.
Grade II tuning for 105E engine (65bhp) or 109E (85bhp) including supply of 2 Weber carburettors & manifolds. Approx.	+ £100. 0. 0.
Grade III tuning for 105E engine (88bhp) including supply of 2 Weber carburettors & manifolds, enlarged & baffled sump and inclusion of oil take-offs for cooler.	+ £256.10. 0.
Grade IV tuning for 105E (100bhp) or 109E (120bhp) as for Grade III	+ £306.10. 0.
Coventry Climax FWA 1100c.c. Stage III engine in place of Ford.	+ £285.15. 0.
Coventry Climax FWE 1220c.c. Stage III engine in place of Ford.	+ £385.15. 0.
B.M.C. "B" Gearbox in place of Ford.	+ £ 13.10. 0.
4-Speed close ratio gears for above (Exchange-unfitted)	+ £ 20. 0. 3.
B.M.C. "A" Gearbox in place of Ford.	+ £ 0. 0. 0.
4-Speed close ratio gears for above (Exchange - fitted)	+ £ 63. 7. 0.
Small Climax flywheel required for above gearbox (Exchange)	+ £ 5. 5. 0.
Borg & Beck 7¼" s.d.p. racing clutch in place of Ford	+ £ 4.15. 0.
Borg & Beck twin plate racing clutch in place of Ford	+ £ 22. 0. 0.
Assembly. (Dependant on specification). Export only.	from £ 70. 0. 0. to £110. 0. 0.
Track testing prior to despatch. Export only.	£ 30. 0. 0.

Optional Extras.

Alternative rear axle gear carrier assys.	£ 26. 0. 0.
Additional fuel tank with cock & fittings.	£ 17.18. 6.
Twin Weber conversion for Climax engines.	£ 54.15. 0.
Oil cooler with fittings.	£ 10. 0. 0.

Price list for the Clubman's Mark 1A. Only one Ford engined car was ever made.

CHAPTER 8

1962 and Later: Who needs an engine at the wrong end anyway?

BY 1962 sports car racing had officially disappeared from international competition. The Manufacturers' Championship was for Grand Touring machines only, although some races ran a new category for 'experimental cars' which looked just like sports racing cars with some interior trim and proper windscreens. The situation in British racing was more relaxed, and most race meetings could be relied upon to cater for sports racing cars. In fact, *Autosport* announced just before Christmas 1961 that the *Autosport* Championship would be open to sports cars in 1962.

Lola Cars were busy producing the new and much more competitive Mark 5 Formula Junior, although the operation of the works FJ cars was handed over to Ron Harris in June to allow the works to concentrate on the F1 Bowmaker-Yeoman Racing Team Mark 4s. Toward the end of the year, they built the first Lola GT, the Mark 6, which, of course, was later to become effectively the forerunner of the Ford GT 40.

Final revisions for the Mark 1

Mark 1s were built only to special order by 1962. The only changes involved wheels and brakes. Finally, the Alfin drums were pensioned off, to be replaced by Girling 9½-inch discs with alloy calipers. Wheels were reduced to 13-inch diameter, as used on the Formula Junior. Three cars were built, one of which was exported to Jesse Wyllie in the US. The two UK cars were sold to Tom Hart, and former Brooklands' motorcycle racer, Bluebelle Gibbs, who ran the car from Len Gibb's Slades Garage at Penn, Buckinghamshire. Bluebelle was a familiar sight at Silverstone club races in the 1960s, as she drove the car on the road to the nearby circuit, often with her over eighty-years-old mother in the passenger seat.

Much loved Bluebelle Gibbs campaigned her car at many Silverstone meetings in the 1960s — usually driving to the nearby circuit on the road.

New rear-engined challengers from Elva and Lotus

The new Elva Mark VI was announced in December 1961. It was a rear-engined car that had a light spaceframe chassis with a stressed aluminium floor. Wheels were 15-inch Elva, although it still used drum brakes. The Coventry Climax FWA engine drove through a modified VW gearbox. Chris Ashmore gave the Elva its debut outing at the 1961 Boxing Day Brands Hatch meeting, where it caused something of a sensation by giving a hard time to Graham Hill's Scuderia Serenissima-entered Ferrari Testa Rossa.

Meantime, over at Cheshunt, Colin Chapman decided it was about time Lotus did something to re-enter the small sports car market too. The Formula Junior 20 had been very successful in 1961, and the rear-engined 23 sports racer was based on this design, together with the 1962 Formula Junior improvements used on the 22. It used a similar, but obviously wider, space-frame, and had the same front suspension. At the rear, the layout was similar to that used on the 1961 F1 car, the 21. Girling 9 1/2-inch discs were fitted. Power initially came from a Cosworth Ford 1098 cc engine: many early 23s were made with this FJ unit, and it was not until 1963 that Lotus introduced the now more familiar twin-cam 23B. Drive to the rear wheels was through either a VW or Renault-based four-speed gearbox/final drive unit. The car was first shown to the public at the 1962 Racing Car Show.

New faces, and some new paint jobs

Most of the existing Lolas changed hands during the close season, with several of them finding their way to the USA to satisfy the still strong North American demand, often via Ian Raby's hands. John Nicholson was set to continue with the ex-Hitches car, and Nairn Ferrier later squeezed an FPF in place of his FWE and added disc brakes to take on allcomers in Scotland and on Northern English circuits, although the car never did go as well with the larger engine. Of the new drivers, the talented Tony Hegbourne was undoubtedly the star with the old 1959 ex-

The talented Tony Hegbourne winning the first event at the BARC Goodwood Members' Meeting on the 24th March 1962 with one of the old 1959 cars. Hegbourne was to be awarded the Brooklands Memorial Trophy for his performance over the season.

Taylor, Gammon, Rees, Mrs Dickinson car. In one of life's small coincidences, Tony Hegbourne had cut his racing teeth in 1959 driving the ex-Richard Utley Tojeiro-Climax: the car that was alongside Eric Broadley and the prototype at the first outing at the back of the Crystal Palace grid in July 1958. Another new owner was Peter Postlethwaite, who bought Laurie Keens' car for the experienced Syd Fox to drive effectively.

Hegbourne showed his class early at the first BARC Goodwood Members' Meeting of 1962 with a win in the first race for up to 1100 cc sports cars. By Easter, the front versus rear-engined battle in the small sports car classes was starting in earnest, and the Lola Mark 1s saw what they were up against. At the Monday Goodwood event, the Sussex Trophy for under 1500 cc sports cars was won by Australian Paul Hawkins in Ian Walker's new Lotus 23 from a closely following John Nicholson. Hegbourne ran his Bill Lacey-tuned car again at the Easter Monday BRSCC Brands, but had to follow Roy Pierpoint's larger Lotus 15.

Nicholson was concentrating on the *Autosport* Championship events, but again could not improve on second place to Paddy Gaston's Elva Mark VI at the Aintree 200 Meeting. Lola fortunes pretty much continued like this throughout the summer, with the cars proving very competitive but usually having to concede narrow defeat to only the best of the Lotus 23's driven by either Hawkins, Mike Beckwith or Rodney Bloor.

Hegbourne managed to set a new 1100 cc sports car lap record of 1 min. 34.2 secs at the 50th Goodwood Members' Meeting on 26th May 1962 in the combined FJ/small sports car race, which was won overall by Hugh Dibley's Lola Mark 3 FJ, just to prove that model was not that bad once it had been developed. But, back with the quickest 23s at the next Goodwood meeting, he could only split Beckwith from ex-Lola man, Peter Boshier-Jones. He had the same result at the BRSCC Oulton Park meeting the next weekend, but beat Beckwith down to third place and this time lost to Bloor by only a coat of paint in an identical race time. Tony lowered the class record to 56.6 secs at the Brands Hatch meeting on 22nd July, but even this was still not quite enough to defeat Beckwith.

Hegbourne finally made it at a BRSCC meeting at Brands Hatch during August, by pressuring Beckwith into a mistake, and causing him to drop to third place behind Laurie Keens' Lotus 23: Beckwith and Hegbourne shared the fastest lap. At this same meeting, the saloon car race was won by a certain Frank Williams in an Austin A40! At the two remaining Goodwood meetings, Hegbourne won the combined FJ/1100 cc sports car race on 1st September but, at the final meeting, three weeks later, he was back looking at Beckwith's exhaust. All of this tremendous effort from Tony in a three-years-old car did not go unrecognised, and he was awarded the 1962 Brooklands Memorial Trophy for his efforts. In 1963 he was to join patron Mike Beckwith to run the very successful Normand team of Lotus 23Bs, and then the less rewarding F2 Cooper T 71s the following year. Sadly this promising driver died in July 1965 from injuries received at the Spa meeting during May of that year.

Elsewhere, Lord Clydesdale started to get the hang of his car as the season wore on. Notable, was a second place at the Snetterton 'Scott Brown Memorial Trophy Meeting' when he split Beckwith and Boshier-Jones 23s, and a win in the sports car race at the MMEC Silverstone event where he beat a Ferrari

Dundee's Nairn Ferrier, with a Coventry Climax FPF engine protruding from the bonnet, is lapped by Innes Ireland's Lotus 19 at the 1962 Gold Cup at Oulton Park.

Oh dear! Racing cars lead a hard life. A rebuild job required for one of the 1959 aluminium bodied cars after this Brands Hatch accident in 1965.

250GT and a Lister-Jaguar. Syd Fox tended to go to the meetings where the quickest Lotuses were and, although he scored some notable places, did not quite match Hegbourne's pace. Nairn Ferrier put his FWE engine to good use at two Rufforth meetings, defeating Lanfranchi's Elva Mark VI each time.

Nicholson nicks the 3 Hours

The *Autosport* Championship was competed for over the year in various sports and GT classes. The five top points scorers in each class qualified for the run-off race at Snetterton, sponsored by the motor sporting magazine at the end of the season. Although John Nicholson was only the third place qualifier in Class E, come the race on 29th September 1962 all started equal, and he outlasted the potentially quicker cars to take a fine win from Dizzy Addicott's Elva and Rodney Bloor's Lotus 23.

Jim fixes it in 1963

Several cars had changed hands over the winter; in particular, Tony Hegbourne's old 1959 Brooklands Trophy winning machine was now with Jim Morley. Tom Hart's 1962 car had gone to Jack Paterson, and Chris Kerrison's earlier car to Ilkeston garage owner, Rex Willoughby.

Lord Clydesdale's car, meanwhile, had gone off with its owner to Singapore. Angus Clydesdale, son and heir to the Duke of Hamilton (family motto: "Never Behind", appropriate for a racing driver!), was serving in the RAF at the time as a Canberra bomber pilot when his Squadron was posted overseas. In Chapter 6 it was noted that there was a thriving racing community in Malaysia, and Clydesdale decided he would ship out the Lola. Almost immediately, at the 1963 Singapore GP, he probably wished he had left the car in Scotland because he had a heavy accident during a very wet practice session. The Lola suffered badly, and was eventually rebuilt by RAF mechanics to look a bit like a Clubman's car with cycle wings, which could be removed for Libre races.

Back in England, Syd Fox ran into his second season with Peter Postlethwaite's car. He continued to carry the fight to the Lotus 23 and Elva Mark VI, and now Mark VII, opposition, but mostly had to settle for the minor places. The highlight was a fine win over his rear-engined rivals at Crystal Palace on 7th September 1963, which was sufficient to promote the Lola to top class qualifier for the *Autosport* Championship. A pair each of 23s and Mark VIIs were the competition. Syd had hopes of repeating John Nicholson's performance in the 3-Hour race, but in the end had to concede victory to Paul Hawkins' Lotus.

Malaysia Grand Prix, 12th April 1965, held in Singapore. The car behind is a Mark 1, although it may not look like it. The local RAF mechanics fitted this clubman's style aluminium body for Lord Clydesdale following an accident. The car in front is also a Lola: in this case, a Mark 2 FJ fitted with a 1650 cc Ford engine, and driven by Dr Tony Goodwin. The two Lolas finished the 180 mile race in third and fourth places.

One noteworthy achievement during 1963 was scored by a Harry O'Brien, who lived close to the Aintree circuit. He briefly owned the ex-Hitches/Nicholson car, and gave it a coat of gold paint. After an outing at the Lancs AC Woodvale Sprint at the end of August, he took the car to the Aintree CC race meeting the following week at the famous venue for the Grand National. At the Liverpool circuit, he squeezed no less than four race wins into the day, which must be some sort of record!

Undoubtedly, the Mark 1 driver of the season was 30-years-old Jim Morley. He made the old 1959 car really fly on occasions. He won first time out at the opening BRSCC Brands Hatch meeting on 23rd March, and again held off the rear-engined brigade at the same venue two months later. Lola had a good day all round on Whit Sunday when Chris Amon drove Reg Parnell's 2.7-litre Lola Climax Mark 4 to a win in the main event, and Morley held off Brian Hart's Terrier-Ford in the sports car race.

Towards the end of the year, Rex Willoughby beat Morley a couple of times, but by this stage he (Morley) had done enough to win the Guards Championship for 1963.

Nothing goes on for ever!

The years were beginning to show at last and, by 1964, not only was the Mark 1's chassis design getting a bit long in the tooth, but the rapid pace of Ford engine development in Formula Junior had also left the Coventry Climax FWA a little breathless too.

Morley, with less success than the year before, and Paterson continued to campaign their cars. Rex Willoughby, reputedly giving Syd Fox's old car an outing at an early March 1964 BRSCC Mallory Park meeting, beat Paterson and moved *Autosport* reporter Paddy McNally to comment:

> "There is no doubt about it, for an old 1100 cc Climax-engined Lola to be able to keep ahead of a twin-cam-engined Lotus 23B says something for Eric Broadley's design ability, and Willougby's driving."

Mid-season Len Gibbs took a win at the traditional AMOC St John Horsfall Silverstone meeting, and Paterson won a couple of classes at BARC Goodwood meetings. Morley and Paterson gave Peter Gethin's Lotus 23 a tough race at the August Bank Holiday Mallory Park meeting with spinner Morley setting fastest lap. For the third year running Morley got a Mark 1 into the *Autosport* 3-Hours at Snetterton, but this was not to be his year either and he spun and damaged the bodywork.

So nearly, in the synchronised swim at Silverstone. March 1965

On 20th March 1965, the BARC ran the Senior Service 200 meeting at the Silverstone GP circuit. The weather was simply awful, and so much so that the race was stopped after 18 of the scheduled 25 laps. When the chequered flag fell, the winner was Jim Clark, who had performed miracles on the slippery surface with the Series 2 Lotus 30. In second place, and actually in the

One of the more successful mid-1960's Mark 1 drivers was Jack Paterson. Here he rounds the sleepers at Tower Crescent at the Whit Monday Crystal Palace meeting on 7th June 1965.

pits at the time the race was stopped, was John Surtees in the very first of the new Lola T70s. Next up, having passed car after car, and not showing any concern with the wet track, was Jack Paterson's 1100 cc Lola Mark 1. If they had kept going for the full distance, who knows?

Editor Gregor Grant reported in *Autosport* the next week:

> "... going well was Jack Paterson in the old red front-engined Lola. He didn't appear to slacken his pace as the conditions worsened. He was gradually catching all but the two leaders."

One of the many casualties of the diabolical conditions was the old prototype, now owned by Bobby Bell. It was his first race with the car, having bought it at the end of 1964, and he ended it backwards in a ditch together with a couple of Lotus 23s, although 600 DKJ was not seriously damaged. He campaigned the car for the remainder of the year and, after a major rebuild in early 1966, notched up four wins, a second and five thirds in twelve races against largely Lotus 23 opposition.

The prototype in early 1966 had gained some ill-fitting later-style bodywork, courtesy of then owner Bobby Bell.

Requiem for Goodwood

The old perimeter track of Westhampnett aerodrome, on the Duke of Richmond and Gordon's Goodwood Estate, opened for business with a JCC/BARC meeting on Saturday 18th September 1948. Something like 15,000 spectators defied the hurdle of post-war petrol rationing to welcome the new circuit. The Duke and Duchess opened the course, and the first event was won by Paul de Pycroft in a re-bodied Jaguar 100.

Lola, of course, had come of age at the 1959 Easter Monday meeting at the circuit, and the history of the Mark 1 model continued to be interwoven with this venue like no other throughout the early 1960s. It was particularly appropriate, therefore, that the Mark 1 should play a part in the final act.

In 1966, the Goodwood Road Racing Company announced that future racing would be limited because of spectator safety concerns, and, soon, the last BARC Members' Meeting was held on 2nd July 1966. The last race of all in the seven race programme was a 5 lap handicap, which was won off the 30 secs mark by Dickie le Strange Metcalfe in the ex-Jack Paterson Lola Mark 1. Equally befitting, Dickie's career had spanned almost the whole Goodwood period too, since he won his first handicap race there seventeen years earlier in 1949 in a Balilla Fiat.

One last hurrah in 1967, courtesy of Mike Fraser

By 1967, Lola Cars Ltd had outgrown the Bromley works and moved to 839 Yeovil Road on the Slough Trading Estate. There, among the T 70s, the T 90 and 92 Indy cars and the T 62 F2s, was effectively a new Mark 1 taking shape.

A young driver called M P R (Mike) Fraser had bought the old Wilkinson/Bekaert car in 1965 and later crashed it. He returned it to the Slough works for a new chassis and a complete update and rebuild. This involved fully rose-jointed suspension, the replacement of drum brakes with discs, plus a short stroke 1149 cc Coventry Climax engine. The rear wheel arches were cut away to house larger tyres, and the whole car was finished in a nice shade of blue.

Fraser concentrated his efforts in 1967 on the Guards Sports Car Championship, which had continuing sponsorship from cigarette manufacturer, Carreras Ltd. Events counting for the Championship were run at Brands Hatch throughout the season, with the last round on Boxing Day, and classes up to and over 1150 cc were accommodated.

Fraser's main Championship rival was Clive Lacey's one-off Brabham BT 8 coupé. This was the ex-Jackie Epstein Targa Florio car, which had been crashed in practice for the Sicilian race. The coupé body had been designed for Epstein, who had unfulfilled plans to run it at Le Mans with an American V8 engine. Another prominent finisher in the larger class that season was Max Mosley with a Meadspeed U2 Mark 6B.

In the smaller class, Fraser often had to give best to the Lotus 23s of John Markey and the former Lola driver, Jack Paterson. However, he scored points in all but two of the twenty-one races he entered throughout the season, and this consistency paid off with a Championship victory over Lacey. Fraser's car was displayed at the 1968 Olympia Racing Car Show, which had replaced the cramped Old Horticultural Hall facilities five years before.

Not very pretty, but effective. Mike Fraser won the 1967 Guards Sports Car Championship.

Successful Lola Mark 1 driver, Jim Morley, seen here at Mallory Park in June 1963.

The Intertech GT

Lord Clydesdale acquired a second Mark 1 chassis on his return to the UK from his spell in Singapore with the RAF. This was fitted with a Chris Steele tuned Ford engine, and run initially with normal bodywork in 1966. Later it was rebuilt as a coupé and used a Diva Valkyr windscreen. It was finished in Clydesdales's dark blue and was named the Intertech GT after his steering wheel business of the same name.

The Intertech GT, pictured at the Nürburgring, was actually a Lola Mark 1 in disguise as a coupé.

Retrospective

Throughout the history of motor racing, there have always been landmark cars that come along to challenge the established competition. Less often, however, does such a vehicle start life as a home-built one-off special, and then, uniquely, go on to be the forerunner of the UK's longest established manufacturer of production competition cars. The Lola Mark 1 has such a distinction. In the words of the *Autosport* reporter after the 1960 Silverstone International Trophy, up to 1500 cc sports car event, the race:

> "....proved the utter and complete superiority of (the) Lola (Mark 1) beyond all doubt."

Postscript

WHEN I started to write this book in early 1997, the intention was to commemorate the fortieth anniversary of both the Lola Mark 1 and the parent Company. At the time, the future of Lola Cars Ltd looked exciting, with another attempt at Formula 1, even if one had concerns about the wisdom of the venture. What a difference a few months can make in the turbulent world of motor sport. Sadly, by April, the Formula 1 venture had collapsed, and the resultant financial burden put Lola Cars Ltd into administration.

By August the future of the Company had become safeguarded when Martin Birrane acquired it. So, although Lola Cars and the Mark 1 will now jointly celebrate their anniversary, the original founder will unfortunately no longer be at the helm. None of the recent happenings, however, takes anything away from the brilliance of design of Eric Broadley's first Coventry Climax-engined sports car.

David Pratley December 1997

Appendix A

Track Tests

'Fabulous Lola'
reproduced from *Autosport*
23 October 1959

'Lola' – Track Report
reproduced from
Sports Cars Illustrated November 1959

These magazine features have, by necessity, needed to be reduced in size from the originals and as such legibility may not be that good.

AUTOSPORT, OCTOBER 23, 1959

FINE THANKS! John Bolster grins happily in the cockpit while Eric Broadley looks just as pleased. The venue is Snetterton.

JOHN BOLSTER TESTS THE

FABULOUS LOLA

The All-conquering 1,100 c.c. Sports Racing Car

No success story has ever been quite so sudden and complete as that of Eric Broadley and his Lolas. It is literally true that the arrival of this little bombshell has completely rejuvenated the 1,100 c.c. sports class, which was rapidly losing the interest of the public. It is equally true to say that Lola victories are now taken for granted, and it would almost be headline news if one were beaten. Thus, no better choice could be made for a road test subject in this important issue of AUTOSPORT.

The basis of the Lola is a multi-tubular frame which weighs just under 60 lb. bare. It is suspended independently all round on what is generally termed the wishbone system. In front, the tubular wishbones have a very wide base and face forward at an appreciable angle, by virtue of the rear bearings on the frame being more widely spaced than the front ones. The stub axles articulate on nylon ball joints at the top and trunnions at the bottom. The steering has an Elektron case, which contains a Lola rack and a Morris Minor pinion.

The independent rear suspension is by helical springs and Armstrong telescopic dampers, and the suspension arms work on rubber bushes. These features are shared with the front suspension, but here the resemblance ends. Immense trouble has been taken to avoid rear wheel steering, and the Elektron housings for the Timken taper roller hub races are located in all planes. Long trailing arms of substantial section look after the fore and aft location, and the large diameter tubular half shafts, with their Hardy Spicer universal points, act as the upper wishbones. The bottom wishbones have wide bases, while in front of them, and working on the same centres, a pair of extra radius arms still further inhibit steering by the rear wheels.

The Elektron bolt-on wheels carry 5.00 x 15 ins. rear tyres, while the front size is 4.50 x 15 ins. The Lockheed brakes with Alfin drums have wider shoes in front. Right in the nose of the car, a Gallay radiator with a copper core keeps the engine cool. A fuel tank is mounted to the left of the scuttle, and an extra one can be fitted on the right for long-distance races. A tap allows either tank to be selected.

A Coventry-Climax engine in Stage 3 tune probably develops 90 b.h.p. with two twin-choke Weber carburetters. This well-known light alloy single overhead camshaft unit drives through a 7¼ ins. Borg and Beck clutch with a racing centre plate. The gearbox casing is from the small B.M.C. box, containing special Lola close-ratio gears. The chassis-mounted hypoid is also a B.M.C. product in a Lola Elektron case. Various ratios are available for different courses.

The Lola which I used for my test was the actual one which won the 1,100 c.c. class in the T.T., and indeed it had not been touched since that epic. I stole it out of the paint shop, where it was due to receive another coat of B.R.G., so the appearance was not as immaculate as usual, but all the performance was there.

On taking the wheel, I found that the lateral location provided by the bucket seat was very good, to the extent that I had positively to force my posterior into position; however, I was comfortable once I had settled into place. The engine started at once, the clutch gripped very positively, and I was away.

The special close-ratio gears make a purposeful scream as one accelerates. Wheelspin can be induced, but it is easy to make a good getaway and the engine does not tend to "fluff". The acceleration is obviously very fine indeed, and continues right up the range. One hundred m.p.h. cames up on any straight, and it is only above this considerable velocity that the Lola ceases to feel like a 3½-litre car.

I am sure that the superiority of the Lola in circuit racing comes from its sheer cornering power. The steering is extremely quick and very direct, but one gets used to this at once. On fast curves, the handling characteristic tends towards understeering. An extremely good feature is the stability over bumps, there being no tendency whatever to hop. It is, above all, the way in which the

THERE IS a good driving position and all the necessary instruments are in the right place. The car even has door pockets!

AUTOSPORT, OCTOBER 23, 1959

THE SLEEK LINES of the Lola are well shown in this photograph, as is Bolster's comfortable driving position. The car was due for a respray when Bolster "stole" it from the works!

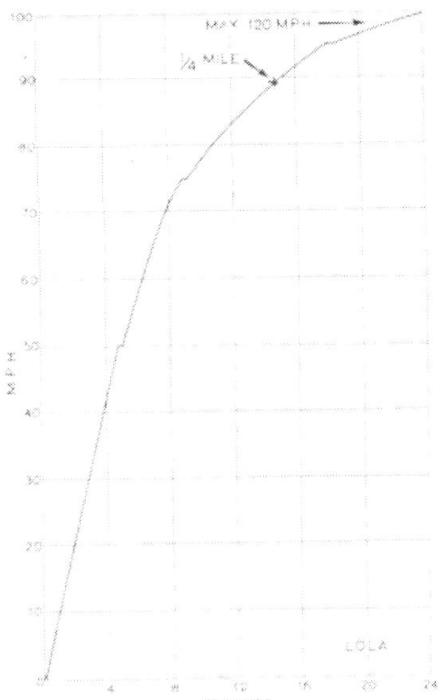

ACCELERATION GRAPH

SPECIFICATION DATA

Car Tested: Lola sports-racing two-seater. Price in kit form, £1,549.

Engine: Four cylinders, 72.4 mm. x 66.6 mm. (1,097 c.c.). Direct valve operation from single overhead camshaft driven by gears and chain. Head and block of light alloy. Compression ratio, 10.5 to 1. Approx. 90 b.h.p. at 7,000 r.p.m. Two twin-choke Weber carburetters. Lucas coil and distributor.

Transmission: Borg and Beck 7¼ ins. single dry plate clutch with racing centre plate. Four-speed gearbox with central lever, ratios 4.55, 5.60, 7.55, and 10.92 to 1. Open propeller shaft. Chassis-mounted hypoid unit. Hardy Spicer articulated half shafts without slip joints.

Chassis: Multi-tubular space frame. Independent front suspension by wishbones, helical springs, and telescopic dampers. Rack and pinion steering. Independent rear suspension by wishbones, trailing arms, helical springs and telescopic dampers. Lockheed brakes in 10 ins. Alfin drums, width 1¾ ins. rear, 2¼ ins. front. Bolt-on Elektron wheels, fitted 4.50 x 15 ins. front and 5.00 x 15 ins. rear tyres.

Equipment: 12-volt lighting and starting. Rev. counter. Ammeter. Oil pressure and temperature and water temperature gauges.

Dimensions: Wheelbase, 7 ft. 1 in. Track, 4 ft. Overall length, 11 ft. 9 ins. Width, 4 ft. 11 ins. Weight, 7¼ cwt.

Performance: Maximum speed, 120 m.p.h. (approx.). Speeds in gears: 3rd 95 m.p.h., 2nd 75 m.p.h., 1st 50 m.p.h. Standing quarter-mile, 14.6 secs. Acceleration: 0-30 m.p.h., 2.8 secs.; 0-50 m.p.h., 4.8 secs.; 0-60 m.p.h., 6.4 secs.; 0-80 m.p.h., 10.6 secs.

Fuel Consumption: In races, 19-20 m.p.g.

rear wheels stay glued to the road that makes the Lola such a fantastic car on corners.

On wet roads it is fairly easy to break the rear end away by a burst of throttle on an appropriate gear, but that high-geared steering remains master of the situation. Naturally, a car of this calibre demands some finesse in handling, but once the knack has been acquired one has absolute confidence in one's ability to control the little machine. The authentic drifting attitude may easily be attained by the appropriate use of the loud pedal.

The brakes are extremely powerful. A slight tendency to pull to the right was no doubt a legacy of many hours of racing in the T.T. The Lola is so small and light that the brakes do not have a very arduous task to perform.

For various reasons, a journey to my usual fast test stretch was impracticable, and the performance figures were not obtained under ideal conditions. The ground was damp, and it actually rained quite hard during some of the tests. The performance figures do not therefore represent the ultimate of which the Lola is capable. Even so, they do give a very fair idea of the potency of this little car. They also show that for sheer traction on slippery roads you can't beat an independent rear end.

I was very impressed by the sturdy construction of the Lola. The reliability of these cars has been exemplary in races, and certainly the machine did everything that I asked of it during my tests. It is small but has none of the flimsiness that has characterized some 1,100 c.c. sports cars. The body is practical, even having such things as door pockets, and a very impressive feature is the good steering lock—a rare virtue among sports-racers. It would appear to need virtually no modification to pass the new and more severe international regulations. It is intended to evolve a sleeker nose shape in the interests of better aerodynamics.

Having watched Lolas winning races, one is naturally all the more interested to drive one. Without doubt it is sheer roadholding, which permits extremely fast cornering, that is the secret, backed up by excellent acceleration in the range where it is most useful. A really sound design in which all the basic requirements are met, coupled with a very high standard of construction, adds up to an ideal sports-racing car that is immensely enjoyable to drive. It made one elderly ex-racing driver feel years younger! The manufacturers are Lola Cars, Ltd., Napier Road, Bromley, Kent.

FRONT. The tubular wishbones have a very wide base and face forward by an appreciable angle. The chassis frame weighs just under 60 lbs. bare!

REAR. Inboard brakes are a feature here. Suspension is by helical springs and Armstrong telescopic dampers and the suspension arms work on rubber bushes.

SPORTS CARS ILLUSTRATED/November 1959

LOLA

TRACK REPORT

by John Christy

► Something that every do-it-yourselfer eventually learns is that although you can drive a screw with an old butter knife and hammer a nail with a brick, any job will be done better and easier if you use the right tool for the task. It's the same thing with motor racing. You can play Tazio Nuvolari around the race course in just about any car, but it is a whole lot more fun, faster and, in many ways easier if you do it in a car meant for racing, one that was built from the outset to excel at just that activity.

This was brought forcibly to mind one hot summer day when we met Lola. Not the choice little red-haired vamp of "Damyankees" fame but a choice little blue racing sports car. The only one of its kind in this country at the time, we had gone to Chicago especially to make the acquaintance.

For those unfamiliar with Class G Modified racing, Lola is the latest English lightweight built around the Coventry Climax 1100 engine and, at this writing, the most successful. The particular one we met has yet to lose a race. In four times out, in the hands of owner, Allan Ross, Lola has four wins.

The first day all we did was pry into Lola's innards with Owner Ross and Eric Broadley, her young English architect. As a spare-time sideline to designing buildings, Broadley last year built himself his own little Lotus-eater. Successful, he built another for a friend. Then another and another. So successful have they been, the building industry may have to wait awhile for Broadley's talents.

Typical of her class of car, Lola has a super-light space frame of welded tubing covered with a smooth aluminum skin whose configuration that looks like nothing so much as a miniature Scarab. Running gear, again typical, is a mixture of Triumph and BMC A-series parts, adapted and modified. The rear end is fully independent with its BMC center section modified to carry a pair of TR-3 10-inch Alfin brake drums inboard, machined down to 1¾ inches wide from 2½. Outboard are a pair of specially cast aluminum hub carriers. The special hubs connect the wheels to unsplined Hardy Spicer drive shafts with universal joints at each end. Location is by triangulated lower wishbones, the doubly U-jointed driveshaft and single trailing radius rods. The Armstrong coil-shock units are "pin-jointed" at both ends and do no locating. A very interesting part of this whole affair is that the rear end can be "tuned" or set for varying breakaway characteristics. Broadley calls them "slow" and "fast" tune. The method is sim-

Right: Control arms are angled forward to handle most braking stresses. Below: Coming off the banking with 5000 rpm on the clock, Lola would touch 7200 by the middle of the front straight. This is tantamount to going from 90 mph to 130 in a quarter of a mile!

Right: Lola's undressed rear shows negative camber of back wheels. Angle is more acute when driver and fuel are loaded aboard. Lower right: Detail of outboard end of control arm and hub casting. Static camber setting can be changed by turning offset pin at bottom of hub.

Above: Designer-builder Broadley shows new owner Allen Ross (right) how the competent Lola driver should fit.

Below: Engine compartment is full but not crowded in the FWA Climax-powered Lola. Broadley disapproved of the lightening holes in the generator case. His thought would have been to turn the case down to half thickness in a lathe. Neat idea.

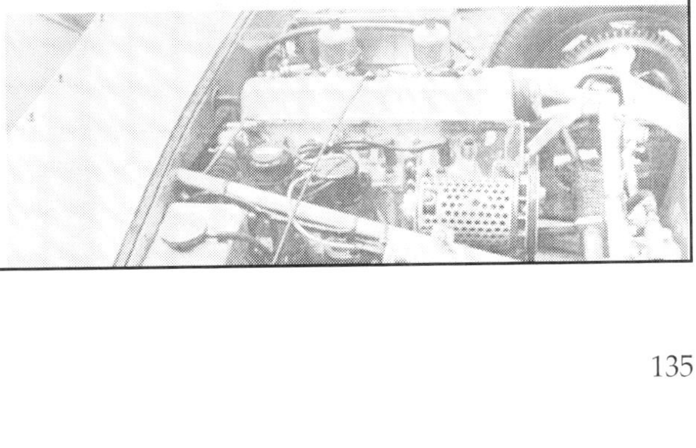

34 SPORTS CARS ILLUSTRATED/NOVEMBER 1959

LOLA Mk III

SPECIFICATIONS

POWER UNIT:
Coventry Climax FWA, Stage III
Four cylinder, water-cooled
Valve Operation....SOHC, valves in line
Bore & Stroke....
2.85x2.62 in (72.4x66.66 mm)
Stroke/Bore ratio 0.92/1
Displacement.........67.1 cu in (1098 cc)
Compression ratio 10.5/1
Carburetion....................Two H-4 S.U.
Max. Power (advertised)
90 bhp @ 7200 rpm
Max. Torque (advertised)
72 lbs-ft @ 5250 rpm
Idle speed:..............................1500 rpm

DRIVE TRAIN:
Gearbox ....MGA
Ratios: I 2.45
II 1.62
III 1.27
IV 1.00
Final Drive 3.72
Other ratios available 4.22, 4.55, 4.88, 5.37
Axle Torque taken by radius rods & wishbones

CHASSIS:
Multi-tube frame, 1 in and ¾ in dia. 20 gauge
Wheelbase ....85 in
Tread, front and rear..........................48 in
Front Suspension..Unequal length tubular wishbones, coil springs
Rear Suspension.....Independent, A-arms and radius rods, coil springs
Shock absorbers..concentric with springs
Steering type.....................rack and pinion
Turns, lock to lock..................................2.0
Brakes ..Al-fin drum, rear brakes inboard
Lining area....................107 square inches
Tire size......4.50/15 front, 5.00/15 rear

GENERAL:
Length ....131½ inches
Height at cowl...........................23½ inches
Clearance ....5 inches
Weight, dry..830 lbs.
Distribution, F/R.....49/51 (with 160 lb. driver)
Fuel capacity...................7 U.S. gallons

Photos: John Christy

ple in action, the inner mounts of the wishbones being moved up or down in a choice of three locations, up being fastest and down being slow. Broadley admits that he's never yet tried it in the fastest setting. Further control is gained by changing to springs with softer or harder rates, the softer the spring, the more "stiction" engendered. Ross's car had the stiffest springs and the low-slow setting since both he and Broadley felt that the faster settings are beneficial only on very high-speed corners. A car thus set-up would be too tricky for tight American courses. Even on the medium setting the car is said to hold until the driver is in far over his head, then letting go in a vicious spin unless extreme skill is exercised. In the slow setting, as we tried it, the car gave ample warning. This allowed cornering with the tail hung out a bit, an impossibility with the fast setting.

In contrast to the rear, the front suspension is perfectly straightforward, using unequal length tubular wishbones with a diagonally mounted coil-and-shock. Steering is by the familiar rack and pinion which gives two turns of the wheel from lock to lock. This item is a mixture of special and standard parts. It has a Morris Minor pinion with a special rack and housing. The track rods are also Morris. The brakes are again Triumph TR-3 Alfins, this time at their full 2½-inch width. Total lining area is 163 square inches. The bolt-on wheels are Cooper-built Elektron castings which carry 4.50 x 15 tires in front and 5.00 x 15 tires at the rear.

Originally, there were two fuel tanks, one on each side of the cowl. Ross, however, has removed the tank on the driver's side since one tank's seven U.S. gallons is sufficient for most American races. When endurance is required, the second tank will be remounted.

Power comes from the usual Coventry Climax 1098 cc engine that is virtually standard wear in this class. Carburetion on this particular "stage 3" engine is by a pair of single-choke 1½-inch SU instruments. The Lola doesn't need the extra poke to stay ahead of the Weber-carbureted Lotuses, Elvas and Coopers. When the extra power becomes necessary, it is there for the price of a pair of twin-choke Webers or SU's.

We arranged to meet the next day at Meadowdale to put the car through its paces. We arrived just as Allan finished warming up the car. He hopped out and motioned us in. This writer's posterior is by no means well-padded but even so the Lola's accommodations are tight. The seat belt felt almost superfluous so snug was the fit. This is all to the good since the driver is held perfectly in place no matter what sort of cornering is being done: there's no strain at all trying to maintain location on a hard turn or series of esses. The wood-rimmed steering wheel is almost vertical, set low, almost too low, and quite far forward requiring a Farina-like straight-arm technique. But in Lola this poses no problem since steering is so effortless.

We set off for an exploratory lap and immediately fell in love with her. Handling was impeccable and friendly. There was a feeling that one would have to be a ham-handed idiot or do something terribly stupid to go very

(Continued on page 77)

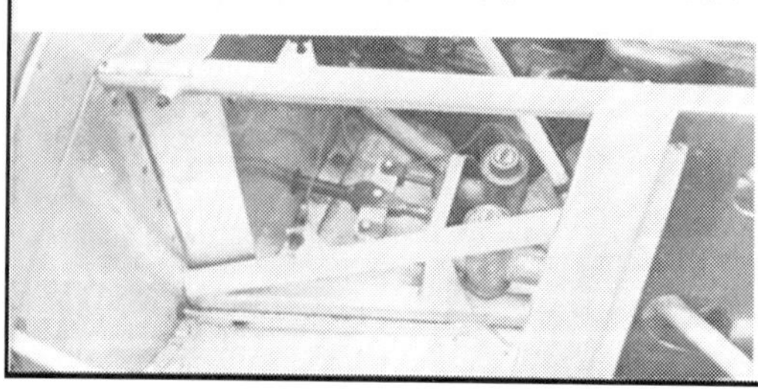

Below: Brakes are actuated by dual master cylinders. By changing position of center push rod more braking can be had at either end.

Below: Lola's cockpit is anything but plush. Layout is fairly comfortable, however. More knee room will be included later.

End of A-arm is set in the slot that gives early breakaway.

Climax engine is canted ten degrees to the left for better balance.

Lola

from page 35

far wrong in this car. Really fast circuitry was out of the question as it had rained the night before and the lower part of the circuit was marred by streaks of dried mud, including just those points where one wants to apply power coming out of a turn. The upper part of the course was clear though, so some appreciation could be gained of the car's fantastic cornering ability. On the banking the only limitation seemed to be the fact that the car bottomed at about 4700 revs or about 90 mph in high gear. We found that we could accelerate off the banking and be hitting 7000 rpm within a quarter of a mile. So effective was the car on the clear portions of the circuit that we and later Ross and Broadley could get around the circuit in fair Lotus and Elva times despite the poor condition of the track lower down.

Both Eric Broadley and Ross had said the car understeered, particularly at lower speeds. It did indeed, thanks to the rear end "tune" and the new, stiffer springs Broadley had brought along, but not to the extent claimed. This is a car that is meant to be driven fast and it should be so driven. At slow or even medium speeds it definitely tends to "push" the front wheels.

Because of warnings from Ross that a spin was a sudden thing and occurred without prior notice we were content at first to drive the car at the point where its feeling of understeering was wholly gone, going no further. Experience with other cars in which early understeer makes a sudden transition, sending the tail out viciously at the last moment made us a bit cautious. We needn't have worried. Broadley's latest springs seem to have cured the sudden breakaway. Toward the end of the day we found ourself whipping around the few clear corners and faster bends with the tail hung safely out in the breeze and without that "oops, Ooops, OOOPS!" feeling that all too many cars engender at such moments. We had the feeling that there was still quite a bit left before things came unstuck and that there would be warning enough before it happened. We also had the humble feeling that we could use much more practice if we were ever to get everything out of the car that the Broadley had designed into it.

One of the nicer aspects of the Lola was the absolute stability of the car at top speed. Some of the English lightweights have a tendency to "hunt" or feel unstable at any speed above 100 mph but not Lola. Where the car is pointed it goes, with no tendency to wander. Things do get a bit light at 7000 rpm but it's nothing one can't cope with.

There is a right tool for every job and Mr. Broadley's Lola seems to be the right one for racing Class G Modified. Anyone with $6000, give or take a few, and a burning desire to run G modified can find Al Ross at any Midwestern SCCA National. His is the blue car that just won. Eric Broadley can be found at 9 Napier Road, Bromley, Kent, England during the week. On weekends he too can be found at the nearest race meeting. His is the silver car that just won. —*jpc*

Appendix B

Individual chassis histories

PROTOTYPE

Registration 600 DKJ.
Colour: Unpainted aluminium.
Engine No. FWA 400-20-7453.
History Built during 1958 in a lock-up garage at Bromley.
Sold to Peter Gammon in late 1958, who later swapped it for BY-3. Sold to George Pitt 1960(?), Bill Mackay / Pit Stop Team 1960, and then on to Peter Borthwick 1962. Driven by Tony Youlton 1964, then sold to Bobby Bell 1964, who replaced the rear of the car from another crashed (unknown) car. Sold to Desmond Freer (1966), and sold to Dick Berger (Milwaukee USA) 1968 who is currently having the car restored by Don Haldenby in the UK.

BY-1

Registration NKT 434.
Colour Initially unpainted aluminium, but by May 1959 Green with White diamond on bonnet.
Engine No. FWA 400-20-8058.
History Built at Byfleet in 1959, with aluminium body.
1959 works car driven by Peter Ashdown. Nürburgring 1000 Kms 1959, retired. 1st in Class and 6th overall Goodwood RAC TT 1959 (co-driven with Allan Ross). Ashdown won BRSCC Sports Car Championship 1959. Sold to Chris Steele at the end of 1959 season, who damaged the aluminium bodywork, and fitted fibreglass for 1960.

To the United States: 1961(?), (Florida USA) sold to Gene Johnson (Memphis USA) ca 1964, then to Bill Niemeyer (Batavia Ohio USA) in 1966. Dick Conrad (Cincinnati USA), and believed repurchased by Niemeyer about 1990.

BY-2

Registration Not registered.
Colour Light Blue.
Engine No. FWA 400-20-8060.
History Built at Byfleet in 1959, with aluminium body.
Fourth 1959 car completed, and the first body built with headfairing. Sold to original US Lola agent, Allan Ross (Chicago USA), and christened 'Lulu'. Sold to Millard Ripley (New York USA) in 1960: 1st SCCA G Mod. Then to Bill Wilbourne a year later and who crashed the car badly at Vineland NJ. Rebuilt by Johnny Boyd (Buffalo NY USA) in 1962 with a virtually new chassis, but retaining some of the original bodywork. Then to Peter D'Amico, Gary Ford (Pipersville Pa USA), Mike Taradash (New York USA), and, via Brian Redman, to Eric Broadley in 1985.

BY-3

Registration FMC 97.
Colour Unpainted aluminium.
Engine No. Not known.
History Built at Byfleet in 1959, with aluminium body.
Supplied initially to Mike Taylor, then swapped with the prototype and driven in most 1959 races by Peter Gammon. Sold to Alan Rees at the end of 1959. To Mrs Brenda Dickinson 1961, then Tony Hegbourne who won the 1962 Brooklands' Memorial Trophy. Jim Morley 1963, won that year's Guards Championship. R Hogarth 1965, and to Brian Williams in 1969. Nigel Moores, 1972, who had it restored by Longbacon Engineering, then John Harper to Alex Seldon, and to Peter Walker. Walker replaced the aluminium body with fibreglass (the original aluminium version still exists in the UK), and won the FIA Historic Sports Car Championship in 1985, then sold the car to Leo Schiewek (Germany).

BY-4

Registration 1133 JN.
Colour White.
Engine No. Not known.
History Built at Byfleet in 1959, with aluminium body.
Supplied to Bernard Cox. Sold to Tommy Hayden in 1960, who later crashed the car and rebuilt it with fibreglass bodywork and a BMC engine. Roy Millbank in 1963, who refitted a Climax engine. To Dr J R Axon in 1966, and George and Jack Rance in 1971. Has been with Rainer Klink (Germany) since 1981.

BR-1

Registration Not registered.
Colour White.
Engine No. Not known.
History Built at Bromley in 1960, with fibreglass body. Supplied to Ron Hathaway (San Jose Ca USA), who repainted the car dark blue. To J Killburn?, then F Tindall—who fitted a Saab engine. Ca 1970 to Gary Ford who refitted a Climax engine. Sold to Bob Sutherland / Bob Seiffert (Boulder Co USA), then to Jim Wolosoff (New Jersey USA) in 1987. Bought by Duncan Dayton (Dover, Mass USA) in 1993, and restored in UK by Sid Hoole in 1995.

BR-2

Registration Not registered
Colour Red.
Engine No. FWA 400-20-8639.
History Built at Bromley in 1960, with fibreglass body. Supplied to Art O'Connor (Lancaster Ca USA). Sold to Art Snyder in 1961, then to Dr William Molle a year later. To J Nunan in 1964, and then bought by David Grossman in 1966. Partly disassembled, and in storage in a warehouse in California for many years. Now in Florida, but still to be reassembled.

BR-3

No record of such a chassis exists.

BR-4

Registration Not known.
Colour Not known.
Engine No. FWA 400-20-8216.
History A car exists in the US with a chassis correctly stamped with a '4' and with a long and verifiable history. The engine is original, and was supplied to Lola in 1960. The car is believed to have come from the UK in 1961 to Matt Kleinmann (Long Island NY USA) then to Jim Adams (Sacramento Ca USA) and Gordon Stewart (Sioux Falls So Dakota USA) from 1963 until 1977. Next owner was John Jeppesen (Minneapolis USA) who sold the car to Cap Chenoweth (Stowe Vt USA) in 1987.

Recently there has been some evidence come to light that suggests this car may have been ex-Peter Boshier-Jones: see BR9.

BR-5

Registration Not known.
Colour Green.
Engine No. Not known.
History Built at Bromley 1960, with fibreglass body. Supplied to David Hitches. Sold to M R J 'Doc' Wyllie (Allison Park Pa USA) in 1960. Sold to Jo Sargent (Raleigh NC USA) in 1963. Bill Barnes bought the car in 1968 (from a Mr Crocker?), and it was burnt out at Atlanta in 1971. Partially restored by Charlie Gibson in 1986, to Don Orosco (Pebble Beach Ca USA), 1996, who is completing the work.

BR-6

Registration SG 16628.
Colour Red/White stripes.
Engine No. FWA 400-20-8503.
History Built at Bromley 1960, with fibreglass body. Supplied to Charles Vögele (Switzerland). 1st in Class 1960 Sebring 12 Hours; 1st in Class 1960 Nürburgring 1000 Kms; Retired with engine failure after 14 hours 1960 Le Mans (in all cases co-driven by Peter Ashdown). 1961 Nürburgring 1000 Kms, 3rd in Class with Ashdown. Next recorded owners were Hansuli Büchi (Switzerland) in 1974, then Karl Bloeche (Switzerland) in 1976 by which time the car was partially restored. Sold to Frank Hilson (Fort Lauderdale Fl USA) via Chris Drake and Keith Duly in 1984. To Herb Wetanson (New York USA) 1992.

BR-7

Registration 2084 JH.
Colour Green.
Engine No. Not known.
History Built at Bromley 1960, with fibreglass body. Supplied to Bill de Selincourt. 2nd in Class, with Chris Lawrence co-driving at 1960 Nürburgring 1000 Kms. Sold to R Waters 1962, then to Chris Kerrison—who reportedly wrote the car off at Mallory Park in July 1963.

Lord Clydesdale (see also BR-21) bought a new chassis from Kerrison in 1966 on his return from Singapore. It is not known whether this is in fact a repaired BR-7 or another different new chassis Kerrison may have had to replace it. This car was run with a Chris Steele Ford 105E engine and normal bodywork during 1966, and later rebuilt as a coupé with Diva Valkyr windscreen and christened the 'Intertech GT', after Clydesdale's steering wheel company.

The car, still in its hardtop form, was sold by David Cottingham to Mike Adlington in 1973, who had the car restored to normal bodywork and a Climax engine by David Vine.

Sold to Richard Loveday in 1977, before passing to Willy Widar (Belgium) a year later, then to Gunther Schultz (Germany).

BR-8

Not identified, but possibly the dark blue car supplied to John Wolfe. Subsequent owners are unknown, although Patrick Behar (France) is reported to have a car claiming this chassis number, but no further history is known.

BR-9

Registration XDW 919.
Colour Green.
Engine No. Not known.
History Built at Bromley 1960, with fibreglass body.
Supplied to Hon Edward Greenall. Sold to Peter Boshier-Jones in 1961 and sold into the USA via Ian Raby. Currently unidentified, but see also BR-4.

BR-10

Registration Not registered.
Colour White.
Engine No. OSCA No 772.
History Built at Bromley 1960, with fibreglass body.
Supplied to Oliver Schmidt (Northbrook Il USA). Won SCCA H Mod Class in 1961. Sold to John Holmes for Guy Bates (Landsdowne Pa USA) to drive. To L Green 1968, and believed to be with Francis Ginther (West Chester Pa USA) since 1985.

BR-11

Registration VBL 858.
Colour Green.
Engine No. Not known.
History Built at Bromley 1960, with fibreglass body.
Supplied to Derek Wilkinson for John Bekaert to drive. Also driven by R Wrenn. Sold to H Balfour in 1964 and Mike Fraser in 1965 after being rolled at Oulton Park. Fraser returned it to the (then) works at Slough for a new chassis and total rebuild. See BR-11/33.

BR-12

Registration VSF 230.
Colour Green.
Engine No. Not known.
History Built at Bromley 1960, with fibreglass body.
Supplied to David and Douglas Bertram in Scotland. Sold to Ralph Miller (Wichita Ka USA). Modified by Miller with wide 13-inch wheels and LSD. Other US owners supposedly Chip Williams, Jim Petric, and in 1971, Gary Schonwald.

Returned to the UK for a David Cottingham rebuild for David Schute. Believed bought by Alex van der Lof (Holland) in 1983, and used by him to win FIA Historic Car Championship in 1986. Via Horst Auer to Gunther Daumler (Germany) in 1994.

BR-13

Registration TOY 914.
Colour Red.
Engine No. FWE 400-30-8228.
History Built at Bromley 1960 as a road car, with thicker fibreglass bodywork and 1220 cc engine. Supplied to Peter Gourlay, who completed over 54,000 trouble free miles. Sold to David Pratley in 1983.

BR-14

Registration Not registered.
Colour Green.
Engine No. Not known.
History Built at Bromley 1960, with fibreglass body.
Supplied to Laurie Keens. Purchased by Peter Postlethwaite/Sid Fox in 1962, sold to Don Jones in 1964 and to Jim Gray in 1965. Via John Pearson and Robert Cooper to Michael Wetherill in 1975 who carried out a major rebuild. Bought by Brian Ashby in 1980, and crashed in a historic race at Silverstone in 1987. Rebuilt 1989.

Dean Butler's - stolen 2004

BR-15

Registration Not registered.
Colour Initially Yellow. Later Red with Gold stripes.
Engine No. Not known.
History Built at Bromley 1960, lightweight fibreglass body.
1960 works car driven by Peter Ashdown. Ashdown won the BRSCC

Sports Car Championship in 1960 with this car. Sold to David McKay (Australia) in 1960. For 1961 McKay added a hardtop, and the car passed to Greg Cusack (Australia), who removed the hardtop in a roll at Catalina Park. Then to John Martin (Australia) in 1963, Frank Demuth (Australia) in 1964 and Tasmanian Bert Howard in 1967. Sold to Kent Patrick (Australia) in 1970, and to Kerry Luckins (Australia) who carried out a restoration in 1974. Sold to Robby Nicholls (Australia) 1984, and to Ian McDonald (Australia) in 1994.

BR-16

Registration Not registered.
Colour Not known.
Engine No. FWA 400-20-8708.
History Built at Bromley 1960, with fibreglass bodywork. Possibly supplied via Grant Clarke (Canada)—see also BR-101—to Hollingshead Co, Ontario, who retailed car care products, and who ran the car briefly as Miss Whiz (or Wiz). Sold to Ross de St Croix (Canada) at the end of 1960. Crashed heavily at the Mosport, Canadian GP meeting in 1962. Rebuilt on a works replacement chassis. Sold to ?, possibly involved in a fire. Remaining history unknown.

The scrapped chassis, aluminium undertray and original chassis plate were sold to George Whipps (Athol Mass USA) who took the remains to Pursang Garage (Peterborough NH USA) intending restoration. Years later, all but forgotten, the bent chassis was taken to the town dump. One of the employees in Pursang Garage kept the undertray (to which the chassis plate was riveted) as a memento, and used it for many years as a furnace heat shield. These remains were discovered in February 1996 by Michael Scully (Peterborough NH USA), who is also searching for the original chassis in the (now closed) town dump.

BR-17

Registration Not registered.
Colour White.
Engine No. Not known, but Climax engine from BY-4 fitted.
History Built at Bromley 1960, with fibreglass body. Supplied to Tommy Hayden (Eire). Sold to Henri Le Roux (Republic of South Africa) in 1962. 3rd overall, and class win in 1963 South African 9 Hours Endurance Race with Gert Coetzer co-driving. Sold to Peter Woolridge (RSA) 1965, who fitted a Ford 105E engine and Alfa Romeo 5 speed gearbox. Sold to John Hatfield (RSA) in 1977. Sold to Lothar

Hoess (Amhurst NH USA): Climax engine refitted. Sold to Dean Trezos (New York USA) in 1986, and to Mark Simpson (Kansas City Mi USA) in 1987. Bought by Stirling Moss in 1989. Completely restored by Don Haldenby with new chassis frame and aluminium bonnet made by Maurice Gomm. Original chassis frame cut up.

BR-18

Registration Not registered.
Colour Green.
Engine No. Possibly FWA 400-20-8495?
History Built at Bromley 1960, with fibreglass body. Supplied to Chris Kerrison. 1st in Class 1961 1000 Kms Nürburgring, with Peter Sargent co-driving. To Rex Willoughby in 1962, and possibly T Barfoot. Sold to Charles Saunders in 1963. Sold to Guido Romani (Italy) in 1993. Chassis and panel work restored by Maurice Gomm, and remaining restoration by Sid Hoole.

BR-19

Registration Not registered.
Colour Green.
Engine No. (FPF 1500 cc Mk 1) Not known.
History Built at Bromley 1960, with heavier gauge tubing and aluminium bodywork. Specially built for Laystall Engineering with twin-cam FPF Coventry Climax engine and 5-speed rear-mounted gearbox made by Laystall/Hewland. Disc brakes. Raced once in the UK by Henry Taylor at the 1961 Goodwood March BARC Members' Meeting to 1st overall in unlimited sports car race. Sold, via Allan Ross, to Bill Stone (Chicago USA) in 1962, and run as SCCA F Modified, sometimes referred to as the 'Lola Medallion'.

Possibly other owners, before Ed Dewell (Broadview Il USA) in 1973. Dewell replaced the damaged FPF, with the engine (FWA 400-20-8060) originally in BY-2. To Dr Terry Bennett (Rochester NH USA) in 1984. Damaged in 1994, when the facilities housing Dr Bennett's collection caught fire.

BR-20

Registration AJD 432.
Colour Blue.
Engine No. Not known.
History Built at Bromley 1960, with fibreglass bodywork. Supplied to John Ewer. Subsequent owners unknown.

BR-21

Registration Not known.
Colour Dark Blue.
Engine No. Not known
History Built at Bromley 1960, with fibreglass body. Supplied to Lord Clydesdale. Shipped to Malaysia in 1963, and crashed in practice for the Singapore GP of that year. Rebuilt with Clubman's style cycle-wing bodywork. Returned to the UK and partially restored with normal fibreglass bodywork in the early 1970s. Unused, until bought by the Earl of March in 1993. Restored by Terry Hall in 1994/95.

BR-22

Registration Not registered.
Colour Red, with Green stripes.
Engine No. FWA 400-20-9304.
History Built at Bromley 1960, with fibreglass body. Exhibited by works at 1961 Racing Car Show. Supplied, via Allan Ross, to Chuck Hall (Dallas Tx USA) in FIA Appendix C specification for 1961 Sebring 12 hours. Later raced by Pete Dawson (Michigan USA). Sold to Jonathan Jenkins (Toledo Oh USA). Possibly the car owned by Tom Thinesen (San Jose Ca USA) since 1983.

BR-23

Registration Not registered.
Colour Black.
Engine No. FWA 400-20-9162.
History Built at Bromley 1960, with fibreglass body. Supplied to Charlie Kurtz (Clearfield Pa USA), via Ross and Millard Ripley, in FIA specification for 1961 Sebring 12 Hours where it won its Class. Sold to Art Tweedale (Arlington Va USA) in 1962. To Phillip Sica (McLean Va USA) in 1972, A J Arlow (Washington DC USA) in 1973, and then Tom Chaffee. Bought by Dick Scott (Pittsburgh Pa USA) in 1979.

BR-24

Registration Not registered.
Colour Originally White.
Engine No. FWE 400-30-9533?
History Built at Bromley 1960/61 with 1220 cc engine, fibreglass bodywork. Supplied to Chan Lye Choon (Malaysia) who won the 1961 Jahore GP. Sold to Brian Stock (Malaysia) in 1963, and

then to Harry Tan (Singapore) in 1966. Repurchased by Brian Stock in 1968 and sold to someone in the UK about 1975, together with a number of spares. It seems the original chassis, with a newer bonnet to replace the partially home-made aluminium/fibreglass Malaysian concoction, was sold via Phillipe Renault (France) to Norbert Michel (France) in 1976.

Allegedly the re-imported spares, together with the chassis plate from the original car, were built up into another car in Northern Europe. This second car was sold through the trade to Joop van Hal (Holland) then Bruce Roberts (Australia) in 1980. Sold to John Dymond (Australia) in 1994.

BR-25

Registration Not registered.
Colour Green.
Engine No. Not known.
History Built at Bromley 1961, with fibreglass bodywork. Supplied to Ian Harrison-Hansley. Subsequent owners unknown.

BR-26

Registration Not registered.
Colour Not known.
Engine No. FWA 400-20-9305.
History Built at Bromley 1961, with fibreglass bodywork. Supplied to Harry Brasewell (USA), driven by Frank Hilson (Ft Lauderdale USA) in 1963. Sold to Capt J R Cavanagh (USA) in 1967. Sold to Hans Glaser (Germany). To Pius Hermann (Germany), and on to Manfred Bode (Germany) since 1980.

BR-27A

Registration Not registered.
Colour White.
Engine No. Not known.
History Built at Bromley 1961, with fibreglass bodywork. Supplied to David Hitches. Crashed and replaced with a second chassis (see BR-27B) before reportedly the original chassis was rebuilt again. Sold to Farnborough Racing in 1961, and initially driven by John Nicholson, who won his class in the 1962 *Autosport* Championship. Also driven by Robin Adair, and B S Glockler. To Harry O'Brien in 1963, and then via Ian Raby to T E Johnston (British Columbia, Canada) in 1965. Sold to ? in 1968 in Edmonton Alberta, and through various other unidentified owners, until finally spending a winter in a snow

bank at Abbotsford BC and changing hands for $350. To Bill Hutton (Clarksville Tn USA) and Raymond Stoutenburg, (Manteo NC USA) in 1976. Sold to Sid Herman (Marblehead Ma USA) approximately 1980.

BR-27B

Possibly a replacement chassis for BR-27. To Nairn Ferrier, Dundee, Scotland. Probably had FWE engine originally, but Ferrier converted to FPF engine, and disc brakes, by 1962. Sold to Blades, Tommy Sleigh in 1963, then to Denis Kingham, Co Down, Northern Ireland, who is reported to have crashed it into a public house! Sold to Mike Fraser in 1964, and F Midgley in 1966. Sold to Ian Holliday in 1970.

BR-28

Registration AZ 5215.
Colour Green?
Engine No. FWE 400-30-9306.
History Built at Bromley 1961, with fibreglass bodywork. Supplied to J Cottle (New Zealand), who won the 1963 NZ Sports Car Championship. Sold to J Riley 1964 and then to W Caldwell in 1965. Sold to Miss I Stephenson (New Zealand) in 1966, and later to Alistair McKee (New Zealand).

BR-29

Registration DB 3229.
Colour Not known.
Engine No. No engine.
History Built at Bromley in 1961 with fibreglass bodywork. Supplied to Doug Lawrence (New Zealand) who fitted a 1500 cc Climax engine from his Cooper Bobtail. Shortly afterwards fitted with a 1500 cc Ford 109E engine. Also driven by Graham Lawrence. Sold to R Dawson and D MacDonald in 1965. Sold to David Oxton (New Zealand) in 1966. Bought by Grant Bloor (New Zealand) in 1974, and to Ian Bisman (New Zealand) in 1975.

BR-30

Registration Not registered.
Colour Not known.
Engine No. Not known.

History Built at Bromley in 1962, with fibreglass bodywork and disc brakes. Supplied to M R J 'Doc' Wyllie (Allison Park Pa USA). 1st G Modified SCCA Nat Championship 1962. Fitted with a Lotus twin-cam engine in 1964. Climax engine fitted, and sold to Lynn Bohlham (Seattle USA). Sold to Gene Deveny (Waco Tx USA) in 1968. Sold to

Gary Ford (Pipersville Pa USA) in 1975, and then to Charlie Gibson (Fairfield Ct USA).

BR-31

Registration 1 UPP.
Colour White.
Engine No. Not known.
History Built at Bromley in 1962, with fibreglass bodywork and disc brakes. Supplied to Bluebelle Gibbs. Sold to David Beckett in 1972, and W James in 1973. In 1975, bought by Mike Ostroumoff, and sold to Count Hubertus Donhoff (Germany) in 1977. Sold to Klaus Scholtyssek (Germany) in 1992.

BR-32

Registration Not registered.
Colour Red.
Engine No. Not known.
History Built at Bromley in 1962, with fibreglass bodywork and disc brakes. Supplied to Tom Hart. Sold to Jack Paterson in late 1962. Sold to Chris 'Dickie' Le Strange Metcalfe in 1965, who won the last race at Goodwood in July 1966 before the circuit was closed for racing. Via John Harper in 1973 to Rupert Glydon.

BR-11/33

Works rebuild of BR-11 at Slough in 1966/7 for Mike Fraser, with new chassis and disc brakes. The car has a 'Slough' chassis plate marked BR-11/33. 1st 1967 Guards Sports Car Championship. Via Norman Lefton to Doug Price in 1969. Via David Cottingham to Robert Cooper in 1973. Sold to American Tony Lapine (in Germany) in 1977, to Sylvain de Cremer (Belgium) and Marc Valvekens (Belgium) in 1990. Sold to Chris Phillips (1997).

BR-101

Registration Not registered.
Colour Not known.
Engine No. M 101.
History Built at Bromley in 1962 as a prototype Mark 1A production sports car, with Martin-Ford 1500 cc engine, disc brakes and steel wheels. Supplied to Grant Clarke (Toronto, Canada). Driven by Gordon Brown in 1963 in Canadian GP. Sold to Frank Salem (Hanover Canada) in 1964, who fitted a BMW 1800TI engine in place of the Ted Martin unit. Sold to Stuart Lamont (Hanover Canada) in 1967. Sold to Bill Hirst (Oshwa Canada) in 1986 who restored the car, and refitted a Ford 1500cc engine.

Other known cars with, at present, incomplete early histories

L-1

A car fitted with a Ford engine in need of major restoration was purchased by Mike Wood in 1972 from a Mr Tattersall of Padiham Lancashire. He (Tattersall) had owned it for about a year, and bought it from 'someone in Chorley Lancashire'. No earlier history is known, although their are some indications it was ex-John Wolfe, Dizzy Addicott (see BR8). The chassis is early, circa 1959, with low mounted master cylinders. Wood restored the car for historic racing in 1976 with a Climax engine. Sold to Chris Draper (Germany) in 1977. Sold to Charlie Gibson (Fairfield Ct USA) 1985, and to Michael Jacobs (Stamford NY USA) in 1994.

L-2

A car came to the USA from Europe in 1962 to David Kiser (Kansas City USA), and was sold on to Bob Hindson and Robert Mc Donnel Jr of the same location later that year. It had an FWE engine, number 400-30-8501, at this stage, and may have originally been white. Sold to Dr Howard Smith (Marlin Tx USA) in 1964, Gene Deveny (Waco Tx USA) in 1968. Bought by Ron van Klein (Long Island NY USA) in 1975. Sold to Harry Reynolds (Pottstown Pa USA) in 1980 and completely restored by him. Sold to Buddy Cooperburg (Brooklyn NY USA), via Charlie Gibson, 1994. Sold to Leigh Keno (New York NY USA) 1997.

L-3

A partially disassembled car was allegedly discovered in a barn on the outskirts of Mexico City in 1993. Bought by Harvey Siegel (New York USA).

L-4

Allegedly a second chassis, complete with corners and disc brakes was purchased by Willy Widar (Belgium) when Patrick Behar (see BR-8) bought his car.

Notes

1. **Coventry Climax Engine Numbers.**

These were in four sections:

(i) The identifying letters (ie FWA, FWB or FWE)
(ii) The type number (all single cam engines were 400)
(iii) Numbers referring to the specification of the engine fittings (eg: with dynamo)
(iv) The actual unit engine number.

2. **'The Thompson Special'.**

A car exists in Australia carrying a Lola badge and bodywork known as the Thompson Special. This car was started in the 1960s by well-known special builder Hedley Thompson for a Mr Cook. It had a chassis similar to a Lola Mark 1, and a Ford engine. Mr Cook was sadly killed racing an Elfin, and the car was never completed at in the time. The project was sold in the early 1970s when Thompson himself died, and was eventually finished during the 1980s.

3. **It is both a brave and foolish man who attempts to catalogue the histories of old racing cars!**

Firstly, chassis plates were not used on most of the Mark 1s. The embryo Lola Cars Ltd in 1959 and 1960 had far too many pressing problems to be overly bothered with the niceties of keeping detailed chassis records. Further, some cars were sold in the UK in partially assembled form to avoid UK purchase tax. Exporting, however, demanded some form of numbering and recognition system. The first number that can be positively identified is the 'BY-2' on the export documents accompanying Ross's car, even though it was actually the fourth 1959 car built. Both BR-1 and 2 had crude hand punched chassis plates attached to the front aluminium shrouding, even though, again, they were most certainly not the first two cars produced at Bromley. It was not until about mid-1960 that the first proper Lola Cars chassis plate appeared, probably attached to BR-16. Some of the earlier UK cars had a hand punched number on the oil filter cruciform support web, just behind the radiator—a fairly vulnerable spot for any front end accident damage.

Dick Samuelson had been a mechanic with the works-supported Fitzwilliam 1960 Formula Junior team. He left at the end of the season to join Lola Cars themselves at Bromley. Dick was the first person to set up a retrospective record system for Mark 1s that had been sold. What appears above uses his initial work as the framework. Much additional research and updating was carried out by George Rance in the mid-1970s, before I took over things in 1984.

As the years have passed it becomes more difficult to unearth the truth. The motoring historian must rely largely on conversations and correspondence with individuals. Memories are unreliable and, regrettably, people are just plain untruthful or forgetful for their own reasons. Sometimes a scrap of information cross-correlates with something someone else has said or written previously; sometimes it confuses and already cloudy picture. The chassis histories recorded in this book are, therefore, my personal opinion at this point in time.

D.A.P. December 1997

Abbreviations

ACF	Automobile Club de France
AMOC	Aston Martin Owners' Club
BARC	British Automobile Racing Club
BET	British Empire Trophy
BMC	British Motor Corporation
BRSCC	British Racing & Sports Car Club
CSI	Commission Sportive Internationale
FIA	Federation Internationale de L'Automobile
JCC	Junior Car Club
MMEC	Midland Motoring Enthusiasts' Club
MMKMC	Maidstone & Mid-Kent Motor Club
NSCC	Nottingham Sports Car Club
SCCA	Sports Car Club of America
SUNBAC	Sutton Coldfield & North Birmingham Automobile Club
USAC	United States Auto Club

Peter Ashdown in the Mark 1 at the International Trophy, Silverstone, 14th May 1960.
Ashdown won the last race on the programme for sports cars up to 1500cc.

Index

Personalities

Index

Cars

(other than Lola)